Advances in Industrial Control

For further volumes:
http//:www.springer.com/series/1412

Zhiqiang Ge · Zhihuan Song

Multivariate Statistical Process Control

Process Monitoring Methods and Applications

Springer

Zhiqiang Ge
Department of Control
Science and Engineering
Institute of Industrial Process Control
Zhejiang University
Hangzhou, Zhejiang,
People's Republic of China

Zhihuan Song
Department of Control
Science and Engineering
Institute of Industrial Process Control
Zhejiang University
Hangzhou, Zhejiang,
People's Republic of China

ISSN 1430-9491 ISSN 2193-1577 (Electronic)
ISBN 978-1-4471-4512-7 ISBN 978-1-4471-4513-4 (eBook)
DOI 10.1007/978-1-4471-4513-4
Springer London Dordrecht Heidelberg New York

Library of Congress Control Number: 2012947424

Printed on acid-free paper

Springer is part of Springer Science+Business Media (www.springer.com)

Series Editors' Foreword

The series *Advances in Industrial Control* aims to report and encourage technology transfer in control engineering. The rapid development of control technology has an impact on all areas of the control discipline. New theory, new controllers, actuators, sensors, new industrial processes, computer methods, new applications, new philosophies…, new challenges. Much of this development work resides in industrial reports, feasibility study papers, and the reports of advanced collaborative projects. The series offers an opportunity for researchers to present an extended exposition of such new work in all aspects of industrial control for wider and rapid dissemination.

Statistical process control (SPC) has now evolved into a group of statistical techniques for monitoring process performance and product quality. An important feature of these methods is that they are used to *monitor* performance and are not a control method *per se* since they contain no automatic feedback mechanism that defines a control action to be taken once a fault condition has been detected.

The classical statistical process control tools are the Shewhart control charts that monitor operational process means and use the assumption of the Gaussian distribution to design upper and lower control limits. The interval of process output variation between these limits defines the normal operating region for the process. If the process output strays outside these limits, this is taken as an indication that the process is operating abnormally and that a process fault or disturbance has occurred. Diagnosis and interpretation of what is causing the process upset is a much more complicated issue and tools like the "cause and effect" chart or a root-cause analysis were developed for this aspect of the performance monitoring problem.

However, the complexity of large-scale industrial operations and the ease with which supervisory control and data acquisition systems and distributed computer control systems can accumulate large quantities of online process data provided an impetus to develop new statistical concepts and tools in the field of SPC. One influential monograph that created interest in extending the techniques of SPC to exploit the information in these industrial datasets was the *Advances in Industrial Control* series monograph *Data-Driven Techniques for Fault Detection and Diagnosis in Chemical Processes* by E. L. Russell, L. H. Chiang, and R. D. Braatz (ISBN 978-1-85233-258-7, 2000).

In the field of feedback control systems, a similar data-driven approach emerged in the late 1980s, but for *controller* performance assessment. The *Advances in Industrial Control* series published two much-cited monographs on these new approaches, namely, *Performance Assessment of Control Loops* by B. Huang and S. L. Shah (ISBN 978-1-85233-639-4, 1999) and *Process Control Performance Assessment* by A. W. Ordys, D. Uduehi, M. A. Johnson (Eds.) (ISBN 978-1-84628-623-0, 2007).

Recent developments in SPC have focussed on procedures that use principal component analysis, partial least squares, independent component analysis, and factor analysis, and these techniques began entering the published literature from the 1990s onward. Now, researchers Zhiqiang Ge and Zhihuan Song have brought together the advanced SPC methods they have developed in this *Advances in Industrial Control* monograph. The volume clearly demonstrates how the methods of SPC have evolved to explore different features of industrial data for performance monitoring. Firstly, the multivariate nature of processes has been absorbed into the field to yield multivariate statistical process control methods. In moving away from the classical assumption of wholly Gaussian variations in the data, Drs. Ge and Song show how to analyse non-Gaussian process situations (Chaps. 3 and 4). The focus of the monograph then moves on to nonlinear process monitoring (Chaps. 5 and 6). To extend the techniques introduced to time-varying processes, the authors introduce adaptive monitoring for nonlinear and multimode processes (Chap. 7). Multimode processes are those that evolve through a range of operating conditions. Monitoring process performance as the process moves through different operating conditions requires methods that minimize the numbers of false alarms and yet are still able to detect faults. The authors devote two chapters to these problems (Chaps. 8 and 9). In Chap. 10, they consider fault detection in vibration signals from an experimental gearbox rig. The penultimate chapter of the monograph is on probabilistic process monitoring (Chap. 11). Plant-wide fault detection and monitoring for large-scale industrial processes have been of considerable interest to the process control community in recent years, and this monograph closes with a chapter that reports the authors' research into using multiblock methods for performance monitoring in such applications.

Throughout the monograph, the authors demonstrate their procedures using some academic examples, along with the quite frequent use of the Tennessee Eastman benchmark chemical process network. Experimental gearbox data is used in the vibration signal analysis and another chapter uses data from a polypropylene production process. Consequently, the reader will find illustrative demonstration examples on a range of industrial processes to study. The monograph should be of interest to engineering and academic readers from the process control and chemical engineering community. Control engineers, industrial statisticians, and graduates from the control community will also find the monograph a valuable contribution to the multivariate statistical process control literature.

Industrial Control Centre
Glasgow
Scotland, UK
2012

Michael J. Grimble
Michael A. Johnson

Preface

Process safety and product quality are two important issues for modern industrial processes. As one of the key technologies in the process system engineering and control area, process monitoring methods can be used to improve product quality and enhance process safety. If a process fault can be anticipated at an early stage and corrected in time, product loss can be greatly reduced. Timely identification of faults can also be used to initiate the removal of out of spec products thereby preserving high standards of product quality. On the other hand, the decisions and expert advice obtained from process monitoring procedures can also be used for process improvement.

In general, process monitoring methods can be divided into three categories: model-based methods, knowledge-based methods, and data-based methods. Model-based methods can be traced to 1970s, at which time they were mainly used in aerospace, engine and power systems. Since model-based methods are based on exact process models, they tend to give more accurate monitoring decisions than the other two method categories, and this is the main advantage of the technique. However, due to the complexity of modern industrial processes, it is very costly to obtain an accurate process model, and in some situations it is not actually possible to develop a process model, per se.

In contrast, knowledge-based methods are often more satisfactory because they are based on the available real-time knowledge of the process behavior and the experience of expert plant operators. Consequently, the monitoring results provided by these methods tend to be more intuitive. However, the creation of the process knowledge base is always a time-consuming and difficult operation requiring the long-term accumulation of expert knowledge and experiences. Although there are limitations to the model-based and knowledge-based methods, they are still popular in particular areas, especially those in which process models can be easily obtained or the process knowledge readily accumulated.

Compared to the model-based and knowledge-based methods, data-driven process monitoring methods have no restrictions on the process model and the associated knowledge and consequently have become more and more popular in recent years. The application areas for the data-driven methods include the important complex industrial processes of the chemical, petrochemical, and biological process industries.

The widespread use of distributed control systems and SCADA technology in the process industries means that large amounts of real-time process data are readily available and this has greatly accelerated the development of data-based process monitoring methods. In addition, the progress made in developing data-mining procedures also provides new technologies for use in process monitoring systems. The data-driven method of Multivariate Statistical Process Control (MSPC) has been the subject of considerable interest from both industry and the academic community as an important tool in the process monitoring area. Three books using Multivariate Statistical Process Control for process monitoring have been published in 1999 (Wang), 2000 (Russell, et al.) and 2001 (Chiang, et al.). Whilst Wang introduced some data-mining technology into MSPC for process monitoring and control, Russell et al. (2000) are mainly concerned with the traditional MSPC methods for process monitoring. Later, Chiang et al. (2001) extended data-based process monitoring to incorporate model-based and knowledge-based methods. It should be noted that the traditional MSPC method is limited to Gaussian, linear, stationary, and single-mode processes. During the last decade, the MSPC-based process monitoring approach has been intensively researched, and many papers have been published. However, to our best knowledge, there is no book on the MSPC-based process monitoring topic that has been published since 2001.

The purpose of this book is to report recent developments of the MSPC method for process monitoring purpose. The book first reviews the recent research works on MSPC and then the shortcomings of the existing approaches are exposed and demonstrated. This provides the motivation for our research and applications work. In our opinion, this book can be assimilated by advanced undergraduates and graduate (Ph.D.) students, as well as industrial and process engineering researchers and practitioners. However, the knowledge of basic multivariate statistical analysis is required, and some familiarity with pattern recognition and machine learning would be helpful.

Acknowledgement

The material presented in this book has been the outcome of several years of research efforts by the authors. Many chapters of the book have gone through several rounds of revisions by discussions with a number of colleagues, collaborators, graduate and Ph.D. students. We would like specifically to thank our colleagues and collaborators, Prof. Youxian Sun for his great supports of the work; Prof. Lei Xie, Prof. Uwe Kruger, Prof. Furong Gao, Prof. Haiqing Wang, Prof. Chunjie Yang, Prof. Jun Liang, Prof. Shuqing Wang, and Prof. Tao Chen who have inspired many discussions in this or related topics over the past years; former and current graduate and Ph. D. students, Dr. Muguang Zhang, Dr. Yingjian Ye, Dr. Kun Chen, Ms. Ruowei Fu, Mr. Zhibo Zhu, Mr. Hong Wang, Mr. Qiaojun Wen, Ms. Aimin Miao, Mr. Le Zhou, and Mr. Hongwei Zhang for their participations in the discussions; and other supporting staffs of the Department of Control Science and Engineering at Zhejiang University. The financial supports from the National Natural Science Foundation of China (NSFC) (60774067, 60974056, 61004134), and the National Project 973 (2012CB720500) are gratefully acknowledged. Last but not least, we would also like to acknowledge Mr. Oliver Jackson (Springer) and other supporting staffs for their editorial comments and detailed examinations of the book.

Contents

1 Introduction 1
1.1 An Overview of This Book 1
1.2 Main Features of This Book 2
1.3 Organization of This Book 3

2 An Overview of Conventional MSPC Methods 5
2.1 Principal Component Analysis 5
2.2 Partial Least Squares 6
2.3 Factor Analysis 7
2.4 Independent Component Analysis 9
2.5 Kernal Principal Component Analysis 10
2.6 Conclusion 11

3 Non-Gaussian Process Monitoring 13
3.1 Introduction 13
3.2 Two-step ICA-PCA Information Extraction Strategy for Process Monitoring 14
3.2.1 Process Monitoring Method 14
3.2.2 TE Benchmark Case Study 17
3.3 Process Monitoring Based on ICA-FA and SVDD 21
3.3.1 Support Vector Data Description 21
3.3.2 Process Monitoring Based on ICA-FA and SVDD 22
3.3.3 Case Study 25
3.4 Conclusions 27

4 Fault Reconstruction and Identification 29
4.1 Introduction 29
4.2 Fault Reconstruction 30
4.2.1 Non-Gaussisan Fault Reconstruction Based on SVDD 30
4.2.2 Fault Diagnosis 32
4.2.3 Simulation Example 33

4.3 Fault Identification . . . 36
4.3.1 PCA-Based Similarity Factor . . . 36
4.3.2 Similarity Factors Based on ICA, FA, and Noise Variance . . 38
4.3.3 Implementation Procedures of Fault Identification Based on Similarity Factors . . . 40
4.3.4 Case Study of TE Benchmark Process . . . 41
4.4 Conclusions . . . 42
4.5 Appendix: Iterative Reconstruction Procedure in the Feature Space 43

5 Nonlinear Process Monitoring: Part 1 . . . 45
5.1 Introduction . . . 45
5.2 Incorporation of Statistical Local Approach and KPCA . . . 46
5.2.1 Statistical Local Approach . . . 46
5.2.2 Introducing Statistical Local Approach into KPCA Monitoring Framework . . . 47
5.3 Nonlinear Process Monitoring Based on Improved KPCA Method 49
5.3.1 Process Monitoring Approach . . . 49
5.3.2 Outline and Implementation . . . 51
5.3.2.1 Normal Operating Condition (NOC) Model Development . . . 51
5.3.2.2 Online Monitoring . . . 52
5.4 Case Studies . . . 52
5.4.1 Numerical Example . . . 52
5.4.2 TE Benchmark Process . . . 55
5.5 Conclusions . . . 58
5.6 Appendix: Proof of Theorem 5.1 . . . 59

6 Nonlinear Process Monitoring: Part 2 . . . 61
6.1 Introduction . . . 61
6.2 Linear Subspace Development . . . 62
6.3 Nonlinear Process Monitoring Based on Linear Subspace Method 65
6.3.1 Process Monitoring Approach . . . 65
6.3.2 Fault Diagnosis . . . 67
6.4 Method Implementation and Remarks . . . 69
6.5 Case Studies . . . 69
6.5.1 Numerical Example . . . 70
6.5.2 TE Benchmark Process . . . 73
6.6 Conclusions . . . 78
6.7 Appendix: Algorithm Complexity Analyses of PCA, KPCA, and BSPCA . . . 79

7 Time-Varying Process Monitoring . . . 81
7.1 Introduction . . . 81
7.2 Local Modeling Strategy Based on JITL and LSSVR . . . 83
7.2.1 LSSVR . . . 83
7.2.2 Local Modeling Approach . . . 84

7.3 Real-Time Problem Considerations 87
7.4 Time-Varying Process Monitoring Based on Local LSSVR Model and ICA-PCA 88
7.5 Simulation Study 90
7.6 Conclusions 94

8 Multimode Process Monitoring: Part 1 95
8.1 Introduction 95
8.2 Mode Clustering and Feature Extraction 96
8.2.1 Fuzzy C-Mean Method for Operation Mode Clustering 97
8.2.2 Two-Step ICA-PCA Feature Extraction for Each Operation Mode 97
8.3 Bayesian-Based Method for Process Monitoring and Fault Identification 98
8.3.1 Multimode Process Monitoring Through the Bayesian Viewpoint 98
8.3.2 Fault Identification and New Operation Mode Determination 101
8.3.3 Remarks 102
8.4 Illustrations and Results 102
8.4.1 A Numerical Example 102
8.4.2 TE Benchmark Case Study 106
8.5 Conclusions 111

9 Multimode Process Monitoring: Part 2 113
9.1 Introduction 113
9.2 Mode Partition and Linear Subspace Construction 114
9.2.1 Data Partition Through Sample Direction 114
9.2.2 Two-Step Variable Selection for Linear Subspace Construction 115
9.3 Two-Dimensional Bayesian Inference for Process Monitoring 117
9.3.1 Method Description and Implementation 117
9.3.2 Algorithm Complexity Analysis 121
9.3.3 Remarks 122
9.4 Case Study of TE Benchmark Process 122
9.5 Conclusions 128

10 Dynamic Process Monitoring 131
10.1 Introduction 131
10.2 Two Dynamical Process Monitoring Methods 132
10.2.1 First Method 132
10.2.2 Second Method 132
10.3 Illustration of the Gearbox System 133
10.3.1 Description of the Gearbox System 133
10.3.2 Analysis of Reference Data 133
10.3.3 Generation of Fault Condition 136

10.4 Overview of Existing Methods ... 136
10.4.1 Independent Component Analysis ... 137
10.4.2 Subspace Model Identification ... 138
10.5 Dynamic Monitoring Schemes for Non-Gaussian Vibration Systems 138
10.5.1 DICA-SVDD Method ... 139
10.5.2 SMI Technique Using the Statistical LA (SMILA) ... 140
10.6 Application Study in a Gearbox System ... 141
10.7 Conclusions ... 146

11 Probabilistic Process Monitoring ... 147
11.1 Introduction ... 147
11.2 PPCA-Based Method for Process Monitoring ... 148
11.3 Bayesian Regularization of PPCA Monitoring Method ... 150
11.4 Multimode Process Monitoring Based on MBPCA ... 152
11.4.1 Mixture Bayesian Regularization of PPCA ... 152
11.4.2 Multimode Process Monitoring ... 154
11.5 Case Studies ... 157
11.5.1 A Numerical Example ... 157
11.5.2 Polypropylene Production Process Application Study ... 160
11.6 Conclusions ... 165
11.7 Appendix: Derivation of the EM Algorithm for the Mixture Bayesian PCA Method ... 166

12 Plant-Wide Process Monitoring: Multiblock Method ... 169
12.1 Introduction ... 169
12.2 Multiblock ICA-PCA (MBICA-PCA) ... 170
12.3 Process Monitoring Based on Two-Level MBICA-PCA ... 171
12.3.1 Fault Detection ... 172
12.3.2 Fault Identification ... 173
12.4 Case Study of TE Benchmark Process ... 175
12.5 Conclusions ... 181

References ... 183

Index ... 191

Notation

$\mathbf{A}$	mixture matrix of the ICA model
a_i	coefficient
$\mathbf{a}$	center of the hypersphere in the SVDD model
$\mathbf{B}$	othogonal matrix of the ICA model
$BIC_{\mathrm{T}^2}(\mathbf{x})$	Bayeaisn inference monitoring statistic based on the T^2
$BIC_{SPE}(\mathbf{x})$	Bayeaisn inference monitoring statistic based on the SPE
$comb\{\cdot\}$	combination operator
CFM	critical fault magnitude
C	tuning parameter in the SVDD model
$\mathbf{c}$	sample mean in the feature space
c_α	normal deviate corresponding to the upper $1-\alpha$ percentile
$\cos\theta$	cosine value of the main angle
$d(\bullet)$	distance calculator
$diag(\cdot)$	make a diagonal matrix
$\mathbf{E}$	residual matrix
$E\{\cdot\}$	expectation
$FDI^s_{T^2}(\mathbf{x})$	fault detection index in each linear subspace under T^2 statistic
$FDI^s_{SPE}(\mathbf{x})$	fault detection index in each linear subspace under SPE statistic
FDI	fault detection index
FDP	fault detection performance
F	feature space
$\mathbf{f}_j$	fault magnitude
GS	Gaussian statistic
h	dimension of feature space
$\mathbf{H}$	bandwidth matrix
H_0	hypothesis test
H_1	hypothesis test
$\mathbf{Ind}(i)$	selected variable index for subspace i
$\mathbf{I}$	identity matrix
I^2	monitoring statistic of the independent component part
I^2_{glob}	global monitoring statistic of the independent component part
$I^2Cont_{block,b}$	block contribution of the I^2 statistic

$I^2Cont_{glob,i}$	global contribution of the I^2 statistic
ICS	independent component subspace
JITL	just-in-time-learning
JP	joint probability
k	number of principal components in the PCA model
$\mathbf{K}$	kernel matrix
KDE	kernel density estimation
$L(\mathbf{P},\beta)$	likelihood function of PPCA
l_{opt}	optimal number of principal factors
$mean(\bullet)$	mean value of the elements
m	variable number
$N(\bullet,\bullet)$	normal distribution
$NFDI_{T^2}(\mathbf{x})$	nonlinear fault detection index under T^2 statistic
$NFDI_{SPE}(\mathbf{x})$	nonlinear fault detection index under SPE statistic
NFDI	nonlinear fault detection index
n	sample number
nc	current sample number
n_{kpca}	number of kernel principal components
NGS	non-Gaussian statistic
$O(\cdot)$	Computational complexity calculator
$\mathbf{P}$	loading matrix of the PCA model
PCS	principal component subspace
$P_{T^2}(N)$	prior probability of the process being normal
$P_{T^2}(F)$	prior probability of the process being abnormal
$P_{T^2}(F\|\mathbf{x})$	fault probability based on the T^2 statistic
$P_{SPE}(F\|\mathbf{x})$	fault probability based on the SPE statistic
$P_{T^2}(\mathbf{x}\|N)$	condition probability under normal situation based on the T^2 statistic
$P_{T^2}(\mathbf{x}\|F)$	condition probability under abnormal situation based on the T^2 statistic
$P_{SPE}(\mathbf{x}\|N)$	condition probability under a situation based on the SPE statistic
$P_{SPE}(\mathbf{x}\|F)$	condition probability under abnormal situation based on the SPE statistic
$P_{I^2}(f\|\mathbf{x}_{fault})$	posterior probability of the faulty sample belongs to a specific faulty mode f under I^2 statistic
$P_{T^2}(f\|\mathbf{x}_{fault})$	posterior probability of the faulty sample belongs to a specific faulty mode f under T^2 statistic
$P_{SPE}(f\|\mathbf{x}_{fault})$	posterior probability of the faulty sample belongs to a specific faulty mode f under SPE statistic
R	radius of the hypersphere in the SVDD model
RBC	reconstruction-based contribution
RS	residual subspace
r	number of independent components in the ICA model
$\mathbf{S}$	independent component matrix of the ICA model
$\mathbf{S}_{mix}$	mixed independent component matrix of the multiblock model

$\mathbf{S}_{\mathrm{global}}$	global independent component matrix of the multiblock model
$\mathbf{s}$	independent component vector
SPE	monitoring statistic of the residual space
$\mathrm{SPE}_{\mathrm{glob}}$	global monitoring statistic of the residual space
$SPECont_{block,b}$	block contribution of the SPE statistic
$SPECont_{glob,i}$	global contribution of the SPE statistic
SPE_{α}	control limit of the SPE monitoring statistic
S_{PCA}	similarity factor of the principal component space
S_{ICA}	similarity factor of the independent component space
S_{FA}	similarity factor of the FA
S_{N}	similarity factor of the noisy part
S_{avg}	weighted average of similarity factors
S_{ICA}	mixed similarity factor
S_{PC_dist}	distance similarity factor of the principal component space
S_{IC_dist}	distance similarity factor of the independent component space
$SubCI_{PCS}(i,j)$	contribution of variable i in subspace j constructed in PCS
$SubCI_{RS}$	contribution of variable i in subspace constructed in RS
$subspace\{\cdot\}$	linear subspace
TDB	two-dimensional Bayesian
$\mathbf{T}$	score matrix of the PCA model
$\mathbf{T}_{\mathrm{mix}}$	mixed score matrix of the multiblock model
$\mathbf{T}_{\mathrm{glob}}$	global score matrix of the multiblock model
$\mathbf{T}_{nl}$	nonlinear latent variable matrix
T_{PCA}	time complexity of PCA
T_{KPCA}	time complexity of KPCA
T_{SPCA}	time complexity of subspace PCA
T_{TDB}	time complexity of two-dimensional Bayesian method
TO_{PCA}	online time complexity of PCA
TO_{KPCA}	online time complexity of KPCA
TO_{SPCA}	online time complexity of subspace PCA
TO_{TDB}	online time complexity of two-dimensional Bayesian method
$Tr(\cdot)$	trace value calculator
$\mathbf{t}$	score vector
T^2	monitoring statistic of the schematic part
T^2_{glob}	Global monitoring statistic of the schematic part
$T^2Cont_{block,b}$	block contribution of the T^2 statistic
$T^2Cont_{glob,i}$	global contribution of the T^2 statistic
$VI(b)$	independent component number extracted in the bth sub-block
$VP(b)$	principal component number extracted in the bth sub-block
$\mathbf{v}$	load vector of KPCA
$\mathbf{W}$	weight matrix of the ICA model
W^c_{ij}	weighted index of each variable in corresponding linear subspace
w	size of moving window

$\mathbf{x}(k)$	sample vector at time point k
$\mathbf{Z}_p$	past arrangement of Hankel matrices
$\mathbf{Z}_f$	future arrangement of Hankel matrices
α	hyperparameter vector in the mixture Bayesian regularization PPCA method
α	significance level
β	reverse of the noise variance
α_i	Lagrange multipliers
$\mathbf{\Phi}$	mapping function
$\mathbf{\Phi}(\cdot)$	nonlinear relationship among the process variables
$\mathbf{\Lambda}$	primary residual
$\mathbf{\Lambda}_n$	primary residual of noisy part
$\mathbf{\Lambda}_s$	primary residual of systematic part
$\sigma_{\max}$	maximum singular value
λ	eigenvalue
$\mathbf{\Theta}$	parameter set
$\mathbf{\Theta}_j$	fault direction
$\mathbf{\Theta}_c^{opt}$	optimal parameter of each BPCA model
θ	process parameter
Σ	monitoring function
Σ_n	monitoring function for noisy part
Σ_s	monitoring function for systematic part
Σ	covariance value
ρ	linear correlation between process variables
ζ	improved residual
ζ_n	improved residual of noisy part
ζ_s	improved residual of systematic part
ξ_i	slack variable in the SVDD model
ξ_j	fault identification index
η_{ICS}	identification index in IC subspace
η_{PCS}	identification index in PC subspace
η_{RS}	identification index in Residual subspace
μ	mean value
χ^2	a Chi distribution

Chapter 1
Introduction

1.1 An Overview of This Book

With the wide use of the distributed control systems in modern industrial processes, a large amount of data has been recorded and collected. How to efficiently use these datasets for process modelling, monitoring and control is of particular interest, as the traditional first-principle model-based method is difficult to use in modern complex processes, which is mainly due to the high human and resource costs or special environments. Different from the first-principle model-based method, the data-based method rarely needs any prior knowledge of the process. By extracting the useful information from the recorded process data, data-based models are also able to model the relationship between different process variables. Particularly, for process monitoring purpose, the multivariable statistical process control (MSPC)-based method has received much attention since the 1990s. The main idea of the MSPC-based monitoring approach is to extract the useful data information from the original dataset, and construct some statistics for monitoring. Most MSPC-based methods can successfully handle the high-dimensional and correlated variables in the process because they are able to reduce the dimension of the process variables and decompose the correlations between them. Therefore, MSPC has become very popular in industrial processes, especially when used for process monitoring.

So far, the most widely used MSPC method for process monitoring may be the principal component analysis (PCA) and the partial least squares (PLS) methods. By extracting the principal components from the process data, and then constructing T^2 and Squared Prediction Error (SPE) statistics for process monitoring, PCA and PLS can both handle the high-dimensional and correlated process variables, and provide detailed monitoring results for each of the data sample in the process. Along past several decades, different improvements and modifications have been made to the traditional PCA and PLS methods. Additionally, some new techniques have also been introduced into the process monitoring area, such as probabilistic PCA, factor analysis, independent component analysis (ICA), kernel PCA, support vector data description (SVDD), etc. most of which were originally proposed in other areas. On the basis of these newly developed and introduced methods, the monitoring performance has been improved for processes under specific conditions. For example,

Z. Ge, Z. Song, *Multivariate Statistical Process Control,* Advances in Industrial Control,
DOI 10.1007/978-1-4471-4513-4_1, © Springer-Verlag London 2013

when the process data are not strictly Gaussian distribution, the traditional PCA method is not sufficient to provide a good monitoring performance. With the introduction of the ICA algorithm, the non-Gaussian data information can be efficiently extracted, on the basis of which new monitoring statistics have been constructed for process monitoring. Nonlinearity is a common data behaviour in many industrial processes; although the conventional MSPC methods fail to describe nonlinear relationship among process variables, some new nonlinear modelling approaches have recently been developed for process monitoring purposes, such as neural network-based MSPC, kernel PCA, linear subspace method, etc. Another common behaviour of modern industrial processes is, their operation condition may be time-varying or they may have multiple operation modes. In these cases, the conventional MSPC methods may not provide satisfactory results. Fortunately, a lot of improvements and new approaches have been developed for monitoring those processes which are time-varying or have multiple operation modes. Some representatives of these methods are adaptive PCA, recursive PLS, moving-window PCA, multimodel method, local model approach, Bayesian inference method, etc. While most of the existing MSPC methods focused on static industrial processes, only a few works have explored the topic of dynamic process monitoring, especially for dynamic non-Gaussian processes. In order to handle the noisy data information in the process, several probabilistic modelling methods have been proposed, e.g. probabilistic PCA and factor analysis. As each process variable is inherently a random variable, the probabilistic model may be more appropriate to describe the relationships among them. Actually, the monitoring performance of most processes has been improved with the introduction of the probabilistic models. Recently, research attention has also focused on plant-wide process monitoring, which is also known as large-scale process monitoring problem in related references. For those processes, the traditional MSPC methods have been extended to the multiblock counterparts, such as multiblock PCA and multiblock PLS. In addition, several new approaches have also been developed on the basis of these multiblock monitoring methods.

The aim of this book is to provide a recent overview of the MSPC method and it introduces some new techniques for the process monitoring purpose. Specifically, this book gives an overview of recently developed methods in different aspects, namely non-Gaussian process monitoring, nonlinear process monitoring, time-varying process monitoring, multimode process monitoring, dynamic process monitoring, probabilistic process monitoring, and plant-wide process monitoring. However, due to the limited space, only some methods have been selected for detailed description in this book.

1.2 Main Features of This Book

The key features of this book are given as follows:

1. According to the complex distribution of given process data, a two-step ICA-PCA based information extraction and process monitoring strategy is introduced. This

method was subsequently improved by using SVDD and factor analysis. For fault diagnosis, the SVDD reconstruction-based non-Gaussian fault diagnosis method is introduced, which can be considered as a complement of reconstruction-based methods for fault diagnosis in the non-Gaussian case. Besides, a similarity-based method is introduced for fault identification.
2. For nonlinear process monitoring, the statistical local approach (LA) is introduced to the traditional kernel PCA modelling structure. This effectively eliminates the restriction of the process data to the Gaussian distribution. Due to the offline modelling and online implementation difficulties of existing methods, a new viewpoint for nonlinear process monitoring, which is based on linear subspace integration and Bayesian inference, is illustrated. Compared to existing nonlinear methods, the new method can both improve the monitoring performance and reduce the algorithm complexity.
3. In order to improve the monitoring performance for time-varying and multimode processes, three new methods are given in detail. First, a local least squares support vector regression-based method is introduced to improve the deficiency of the traditional recursive method for time-varying processes. This greatly enhances the real-time performance for monitoring purpose. Second, a Bayesian inference method is introduced for multimode process monitoring. Third, a two-dimensional Bayesian-based method, which greatly alleviates the lean of the monitoring method to process knowledge and experiences, is also introduced for monitoring nonlinear multimode processes.
4. Very few works have been reported on process monitoring for non-Gaussian dynamic processes. A new monitoring method is introduced in this book for these special processes, which is based on subspace model identification and LA. In contrast to other methods, the new method is more efficient in monitoring non-Gaussian dynamic processes.
5. While the probabilistic PCA method has recently been introduced for process monitoring, it has an inherent limitation that it cannot determine the effective dimensionality of latent variables. A Bayesian treatment of the PCA method is developed and introduced. For multimode process monitoring, this Bayesian regularization method is extended to its mixture form.
6. Based on the traditional multiblock method that has been proposed for plant-wide process monitoring, a two-level MultiBlock ICA-PCA method is introduced. Through this method, the process monitoring task can be reduced and the interpretation of the process can be made more efficiently. Detailed descriptions of fault detection, fault reconstruction and fault diagnosis tasks are provided.

1.3 Organization of This Book

The chapter organization of this book is detailed as follows.

Chapter 1 gives an overview of this book, and illustrates the main features of this book.

Chapter 2 introduces some conventionally used MSPC methods that have been widely used and developed in the past years.

Chapter 3 gives introductions of several non-Gaussian process monitoring methods, including the two-step ICA-PCA strategy, ICA-factor analysis strategy, and the SVDD algorithm.

Chapter 4 introduces the SVDD-based non-Gaussian fault reconstruction and diagnosis algorithm, and the similarity-based fault identification method.

Chapters 5 and 6 introduce the nonlinear process monitoring topic through two different viewpoints.

Chapter 7 gives a detailed description of the local model-based approach for time-varying process monitoring.

Chapters 8 and 9 are two parts which are dedicated on the topic of multimode process monitoring.

Chapter 10 focuses on dynamic process monitoring, which specifically introduces two methods for monitoring non-Gaussian dynamic processes.

Chapter 11 introduces a Bayesian regularization of the probabilistic PCA method and its mixture form, on the basis of which a probabilistic multimode process monitoring approach is also demonstrated.

Chapter 12 gives an introduction of a two-level multiblock method for plant-wide process monitoring.

Some of the materials presented in this book have been published in academic journals by the authors, and are included after necessary modification and updates to ensure accuracy and coherence.

Chapter 2
An Overview of Conventional MSPC Methods

2.1 Principal Component Analysis

Consider an $n \times m$ data matrix $\mathbf{X}$, for m process variables with n measurements of each variable. Here, we assume that the data for each variable have been scaled to zero mean and unit variance. The covariance matrix of $\mathbf{X}$ is defined as

$$\sum = \frac{\mathbf{X}^T \cdot \mathbf{X}}{n-1} \tag{2.1}$$

Let $\lambda_i (i = 1, 2, \ldots, m)$ be the eigenvalues of the matrix of $\sum$ which are arranged in descending order to determine the principal components (PCs), and with the principal component loadings $\mathbf{p}_i$ to be their corresponding eigenvectors. Then, the first k PCs are selected to build the PCA model, so the data matrix $\mathbf{X}$ can be expanded using principal component loadings $\mathbf{p}_i$, score vectors $\mathbf{t}_i$, and a residual matrix $\mathbf{E}$ (Chiang et al. 2001; Qin 2003)

$$\mathbf{X} = \sum_{i=1}^{k} \mathbf{t}_i \mathbf{p}_i^T + \mathbf{E} \tag{2.2}$$

For a new observation $\mathbf{x}_{new}$, the prediction of the PCA model is given by

$$\hat{\mathbf{x}}_{new} = \mathbf{t}_{new} \mathbf{P}^T = \mathbf{x}_{new} \mathbf{P} \mathbf{P}^T \tag{2.3}$$

where $\mathbf{t}_{new}$, is the score vector. The resulting residual is defined as

$$\mathbf{e}_{new} = \mathbf{x}_{new} - \hat{\mathbf{x}}_{new} \tag{2.4}$$

For monitoring purpose, two statistical variables T^2 and SPE can be calculated based on $\mathbf{t}_{new}$ and $\mathbf{e}_{new}$, respectively,

$$T_{new}^2 = \mathbf{t}_{new} \Lambda \mathbf{t}_{new}^T \tag{2.5}$$

$$SPE_{new} = \mathbf{e}_{new} \mathbf{e}_{new}^T \tag{2.6}$$

Z. Ge, Z. Song, *Multivariate Statistical Process Control,* Advances in Industrial Control,
DOI 10.1007/978-1-4471-4513-4_2, © Springer-Verlag London 2013

where Λ is the eigenvalue matrix of the PCA model. The confidence limit of the T^2 and SPE monitoring statistics are determined as follows (Chiang et al. 2001)

$$T^2 \le T^2_{\lim} = \frac{k(n-1)}{n-k} F_{k,(n-k),\alpha} \tag{2.7}$$

$$\begin{aligned} SPE \le SPE_{\lim} &= g\chi^2_{h,\alpha} \\ g &= v/(2m) \\ h &= 2m^2/v \end{aligned} \tag{2.8}$$

where k is the number of PCs, α is significance level, $\mathbf{c}_\alpha$ is the normal deviate corresponding to the upper $1-\alpha$ percentile, m and v are the mean and variance values of SPE of the training dataset. When the statistic values of the new data sample exceed their corresponding control limits, $T^2_{new} > T^2_{\lim}$ or $SPE_{new} > SPE_{\lim}$, a fault is considered to be detected, and further actions should be taken to get the process back in its good condition.

2.2 Partial Least Squares

The principle of the PLS method is similar to that of PCA, except that the PLS method incorporates the information of the quality variables in the process. Given a pair of process and quality dataset $\{\mathbf{X}, \mathbf{Y}\}$, PLS intends to decompose $\mathbf{X}$ and $\mathbf{Y}$ into a combination of scores matrix $\mathbf{T}$, loading matrices $\mathbf{P}$ and $\mathbf{Q}$, and weight matrix $\mathbf{W}$. The relationship between $\mathbf{X}$ and $\mathbf{Y}$ can be described by the following equations

$$\mathbf{X} = \mathbf{T}^T\mathbf{P} + \mathbf{E} \tag{2.9}$$

$$\mathbf{Y} = \mathbf{T}\mathbf{Q}^T + \mathbf{F} \tag{2.10}$$

The regression matrix of the PLS model between the process and quality variables can be determined as follows

$$\mathbf{R} = \mathbf{W}(\mathbf{P}^T\mathbf{W})^{-1}\mathbf{Q}^T \tag{2.11}$$

Given a new process data sample, $(\mathbf{x}_{new}, \mathbf{y}_{new})$, the principal components are calculated in the first step, which is given as

$$\mathbf{t}_{new} = \mathbf{x}_{new}\mathbf{W}(\mathbf{P}^T\mathbf{W})^{-1} \tag{2.12}$$

In the next step, the T^2 and SPE monitoring statistics can be constructed for process monitoring. Different from the PCA approach, a total of four monitoring statistics can be used for processs monitoring in the PLS model, two of which correspond to the data information of the quality variables. Detailed description of the PLS model and its monitoring strategy can be found in Chiang et al. (2001).

2.3 Factor Analysis

Similar to the probabilistic PCA method, factor analysis concentrates on the latent variables $\mathbf{t}$ whose distribution are Gaussian, and the original measurement variables $\mathbf{x}$ are treated as linear combination of $\mathbf{t}$ plus small additive white noise $\mathbf{e}$. The aim of FA is to find the most probable parameter set $\Theta = \{\mathbf{P}, \Sigma\}$ in the model structure, which is described as follows (Bishop 2006)

$$\mathbf{x} = \mathbf{P}\mathbf{t} + \mathbf{e} \tag{2.13}$$

where $\mathbf{P} = [p_1, p_2, \ldots, p_l] \in \mathbf{R}^{m \times l}$ is the loading matrix, just as that in PCA. The variances matrix of measurement noise $\mathbf{e}$ is represented by $\Sigma = diag\{\lambda_i\}_{1,2,\ldots,m}$, in which different noise levels of measurement variables are assumed. If all the λ are assumed to be of the same value, then FA is equivalent to PPCA. If all λ_i, $i = 1, 2, \ldots, m$ are assumed to be zero, then FA becomes PCA. Therefore, FA is the general description of Gaussian latent model structure. PPCA and PCA are just two special cases of FA.

In the FA model, the latent variable $\mathbf{t}$ is assumed to be Gaussian distribution with zero mean and unity variance, which is $\mathbf{t} \in N(\mathbf{0}, \mathbf{I})$, where $\mathbf{I}$ is the unity matrix with appropriate dimension. Precisely, the distribution of the latent variable $\mathbf{t}$ is

$$p(\mathbf{t}) = (2\pi)^{-k/2} \exp\left(\frac{1}{2}\mathbf{t}^T\mathbf{t}\right) \tag{2.14}$$

Then, the conditioned probability of the measured process variable $\mathbf{x}$ is given as

$$p(\mathbf{x}|\mathbf{t}) = (2\pi)^{-m/2}|\Sigma_{\mathbf{e}}|^{-1/2} \exp\left\{-\frac{1}{2}(\mathbf{x} - \mathbf{P}\mathbf{t})^T \Sigma_{\mathbf{e}}^{-1}(\mathbf{x} - \mathbf{P}\mathbf{t})\right\} \tag{2.15}$$

Based on Eqs. (2.14) and (2.15), the probability of $\mathbf{x}$ can be calculated as

$$p(\mathbf{x}) = \int p(\mathbf{x}|\mathbf{t})p(\mathbf{t})d\mathbf{t} = (2\pi)^{-m/2}|\mathbf{C}|^{-1/2} \exp\left(-\frac{1}{2}\mathbf{x}^T\mathbf{C}^{-1}\mathbf{x}\right) \tag{2.16}$$

where $\mathbf{C} = \Sigma_{\mathbf{e}} + \mathbf{P}\mathbf{P}^T$ is the variance matrix of the process data, thus the distribution of $\mathbf{x}$ can be represented as $\mathbf{x} \in N(\mathbf{0}, \Sigma_{\mathbf{e}} + \mathbf{P}\mathbf{P}^T)$, $|\mathbf{C}|$ is to calculate the discriminant value for the $\mathbf{C}$.

The parameter set $\Theta = \{\mathbf{P}, \Sigma_{\mathbf{e}}\}$ can be estimated by the expectation and maximization (EM) algorithm, which is an iterative likelihood maximization algorithm. EM is a widely used algorithm for parameter estimation due to its simplicity and efficiency. Besides, it can also handle incomplete datasets, which is also very attractive to the application in process industry. The EM algorithm can be partitioned into two steps: the E-step and the M-step. The maximum likelihood function of the process data can be given as

$$L = \sum_{i=1}^{n} \ln\{p(\mathbf{x}_i, \mathbf{t}_i)\} = \sum_{i=1}^{n} \ln\{p(\mathbf{t}_i|\mathbf{x}_i)p(\mathbf{x}_i)\} \tag{2.17}$$

where the posterior distribution of each latent variable can be calculated as

$$p(\mathbf{t}_i|\mathbf{x}_i) = (2\pi)^{-k/2}|\mathbf{M}|^{-1/2}\exp\left\{-\frac{1}{2}(\mathbf{t}_i - \mathbf{Q}\mathbf{x}_i)^T\mathbf{M}^{-1}(\mathbf{t}_i - \mathbf{Q}\mathbf{x}_i)\right\} \tag{2.18}$$

where $\mathbf{Q} = \mathbf{P}^T(\mathbf{P}\mathbf{P}^T + \Sigma_{\mathbf{e}})^{-1}$, $\mathbf{M}^{-1} = \mathbf{I} - \mathbf{Q}\mathbf{P}$. In the E-step of the EM algorithm, the expectation value of the likelihood function L is calculated as follows, which is based on the first and second statistics of the latent variable

$$E(L) = -\sum_{i=1}^{n}\left\{\frac{1}{2}\mathbf{x}_i^T\Sigma_{\mathbf{e}}^{-1}\mathbf{x}_i - \mathbf{x}_i^T\Sigma_{\mathbf{e}}^{-1}\mathbf{P}E(\mathbf{t}_i|\mathbf{x}_i) + \frac{1}{2}tr[\mathbf{P}\Sigma_{\mathbf{e}}^{-1}\mathbf{P}E(\mathbf{t}_i\mathbf{t}_i^T|\mathbf{x}_i)]\right\}$$
$$-(n/2)\ln\Sigma_{\mathbf{e}} + cont \tag{2.19}$$

where $E(\cdot)$ and $tr(\cdot)$ are expectation and trace calculators, *cont* represents a constant value, and the first and second statistics of the latent variables are given as follows

$$\hat{\mathbf{t}}_i = \mathrm{E}\{\mathbf{t}_i|\mathbf{x}_i\} = \mathbf{Q}\mathbf{x}_i \tag{2.20}$$

$$\mathrm{E}\{\mathbf{t}_i\mathbf{t}_i^T|\mathbf{x}_i\} = \mathbf{I} - \mathbf{Q}\mathbf{P} + \mathbf{Q}\mathbf{x}_i\mathbf{x}_i^T\mathbf{Q}^T \tag{2.21}$$

In the M-step of the EM algorithm, by maximizing the expectation value of the likelihood function, taking the partial derivatives of L with respect to the parameter set $\Theta = \{\mathbf{P},\ \Sigma_{\mathbf{e}}\}$, and setting them to zero, the optimal values of the two parameters can be determined, which are given as

$$\mathbf{P} = \left(\sum_{i=1}^{n}\mathbf{x}_i\hat{\mathbf{t}}_i^T\right)\left(\sum_{i=1}^{n}\mathrm{E}\{\mathbf{t}_i\mathbf{t}_i^T|\mathbf{x}_i\}\right)^{-1} \tag{2.22}$$

$$\sum_{\mathbf{e}} = \frac{\sum_{i=1}^{n} diag\left(\mathbf{x}_i\mathbf{x}_i^T - \mathbf{P}\hat{\mathbf{t}}_i\mathbf{x}_i^T\right)}{n} \tag{2.23}$$

By calculating the E-step and M-step iteratively, the final optimal parameters for the FA model can be obtained.

For process monitoring, the T^2 and *SPE* monitoring statistics based on the FA model can be constructed as follows

$$T_i^2 = \left\|\hat{\mathbf{t}}_i\right\|^2 = \mathbf{x}_i^T\mathbf{Q}^T\mathbf{Q}\mathbf{x}_i \le \chi^2_{\alpha,k} \tag{2.24}$$

$$\hat{\mathbf{e}}_i = \mathrm{E}\{\mathbf{e}_i|\mathbf{x}_i\} = (\mathbf{I} - \mathbf{P}\mathbf{Q})\mathbf{x}_i \tag{2.25}$$

$$SPE_i = \left\|\Sigma_e^{-1/2}\mathbf{e}_i\right\|^2 = \mathbf{x}_i^T(\mathbf{I} - \mathbf{P}\mathbf{Q})^T\Sigma_e^{-1}(\mathbf{I} - \mathbf{P}\mathbf{Q})\mathbf{x}_i \le \chi^2_{\alpha,m} \tag{2.26}$$

2.4 Independent Component Analysis

Independent component analysis (ICA) is a statistical and computational technique for revealing hidden factors that underlie sets of random variables, measurements, or signals. ICA was originally proposed to solve the blind source separation problem. To introduce the ICA algorithm, it is assumed that m measured variables, $\mathbf{x}(k) = [x_1(k), x_2(k), \ldots, x_m(k)]$ at sample k can be expressed as linear combinations of $r(\leq m)$unknown independent components $[s_1, s_2, \ldots, s_r]^T$, the relationship between them is given by Hyvarinen and Oja (2000)

$$\mathbf{X}^T = \mathbf{AS} + \mathbf{E} \tag{2.27}$$

where n is the number of measurements, $\mathbf{X} \in \mathbf{R}^{n \times m}$ is the data matrix, $\mathbf{A} = [\mathbf{a}_1, \mathbf{a}_2, \ldots, \mathbf{a}_r] \in \mathbf{R}^{m \times r}$ is the mixing matrix, $\mathbf{S} = [s_1, s_2, \ldots, s_r] \in \mathbf{R}^{r \times n}$ is the independent component matrix, and $\mathbf{E} \in \mathbf{R}^{l \times n}$ is the residual matrix. The basic problem of ICA is to estimate the original component $\mathbf{S}$ and the mixing matrix $\mathbf{A}$ with $\mathbf{X}$, therefore, the objective of ICA is to calculate a separating matrix $\mathbf{W}$ so that the components of the reconstructed data matrix $\hat{\mathbf{S}}$ become as independent of each other as possible, given as

$$\hat{\mathbf{S}} = \mathbf{W}\mathbf{X}^T \tag{2.28}$$

Before applying the ICA algorithm, the data matrix $\mathbf{X}$ should be whitened, in order to eliminate the cross-correlations among random variables. One popular method for whitening is to use the eigenvalue decomposition, considering $\mathbf{x}(k)$ with its covariance $\mathbf{R}_x = \mathbf{E}\{\mathbf{x}(k)\mathbf{x}(k)^T\}$, the eigenvalue decomposition of $\mathbf{R}_x$ is given by

$$\mathbf{R}_x = \mathbf{U}\Lambda\mathbf{U}^T \tag{2.29}$$

The whitening transformation is expressed as

$$\mathbf{z}(k) = \mathbf{Q}\mathbf{x}(k) \tag{2.30}$$

where $\mathbf{Q} = \Lambda^{-1/2}\mathbf{U}^T$. One can easily verify that $\mathbf{R}_z = E\{\mathbf{z}(k)\mathbf{z}(k)^T\}$ is the identity matrix under this transformation. After the whitening transformation, we have

$$\mathbf{z}(k) = \mathbf{Q}\mathbf{x}(k) = \mathbf{QA}\mathbf{s}(k) = \mathbf{B}\mathbf{s}(k) \tag{2.31}$$

where $\mathbf{B}$ is an orthogonal matrix, giving the verification as follows

$$E\{\mathbf{z}(k)\mathbf{z}(k)^T\} = \mathbf{B}E\{\mathbf{s}(k)\mathbf{s}(k)^T\}\mathbf{B}^T = \mathbf{BB}^T = \mathbf{I} \tag{2.32}$$

Therefore, we have reduced the problem of finding an arbitrary full-rank matrix $\mathbf{A}$ to the simple problem of finding an orthogonal matrix $\mathbf{B}$. Then, we can estimate as follows

$$\hat{\mathbf{s}}(k) = \mathbf{B}^T\mathbf{z}(k) = \mathbf{B}^T\mathbf{Q}\mathbf{x}(k) \tag{2.33}$$

From Eqs. (2.28) and (2.33), we can get the relation between **W** and **B**

$$\mathbf{W} = \mathbf{B}^T\mathbf{Q} \tag{2.34}$$

There are two classic measures of non-Gaussianity: kurtosis and negentropy (Hyvarinen and Oja 2000). Kurtosis is the fourth-order cumulant of a random variable, unfortunately, it is sensitive to outliers. On the other hand, negentropy is based on the information-theoretic quantity of differential entropy. When we have obtained the approximate forms for negentropy, Hyvarinen (1997) introduced a very simple and efficient fixed-point algorithm for ICA. This algorithm calculates the independent components one by one through iterative steps. After all vectors $\mathbf{b}_i$ $(i = 1, \ldots, m)$ have been calculated and put together to form the orthogonal mixing matrix **B**, then we can obtain $\hat{s}(k)$ and demixing matrix **W** form Eqs. (2.33) and (2.34), respectively.

Dimension reduction of ICA is based on the idea that these measured variables are the mixture of some independent component variables. The performance and interpretation of ICA monitoring depend on the correct choice of the ordering and dimension of the ICA model. Because negentropy is used for measurement of non-Gaussianity, we can select suitable number of ICs by checking their negentropy values. If the negentropy value of current IC is zero or approximately zero, it indicates that the non-Gaussian information of the process has already been extracted, and the rest of the process is Gaussian, which can be analyzed by conventional MSPC methods such as PCA.

Based on the identified ICA model, two monitoring statistics can be constructed for monitoring, which are defined as I^2 and *SPE*, given as (Lee et al. 2004; Ge and Song 2007)

$$I^2 = \mathbf{s}^T\mathbf{s} \tag{2.35}$$

$$SPE = \mathbf{e}^T\mathbf{e} \tag{2.36}$$

where **s** and **e** are independent components and residuals, respectively.

2.5 Kernal Principal Component Analysis

As a simple linear transformation technique, PCA compresses high-dimensional data into low dimension with minimum loss of data information. When the algorithm is carried out in the feature space, KPCA is obtained. This algorithm was originally proposed by Schölkopf et al. (1998). Suppose the original training dataset is $\mathbf{x}_1, \mathbf{x}_2, \ldots, \mathbf{x}_n \in R^m$, where n is the sample number and m is the number of process variables. The feature space is constructed by using a nonlinear mapping: $R^m \xrightarrow{\Phi(\cdot)} F^h$, where $\Phi(\cdot)$ is a nonlinear mapping function and h is the dimension in feature space, which is assumed to be a large value. Similar to PCA, the covariance matrix in the feature space F can be calculated as

$$\sum_F = \frac{1}{n}\sum_{i=1}^{n}[\Phi(\mathbf{x}_i) - c][\Phi(\mathbf{x}_i) - c]^T \tag{2.37}$$

where c is the sample mean in the feature space. Conveniently, denote $\bar{\Phi}(\mathbf{x}_i) = \Phi(\mathbf{x}_i) - c$ as the centered feature space sample. The kernel principal component can be obtained by the eigenvalue problem below

$$\lambda \mathbf{v} = \sum_F \mathbf{v} = \frac{1}{n} \sum_{i=1}^{n} [\bar{\Phi}(\mathbf{x}_i)^T \mathbf{v}] \bar{\Phi}(\mathbf{x}_i) \tag{2.38}$$

$$\mathbf{v} = \sum_{i=1}^{n} \alpha_i \bar{\Phi}(\mathbf{x}_i) \tag{2.39}$$

where λ and $\mathbf{v}$ denote eigenvalue and eigenvector of the covariance matrix Σ_F, respectively. The α_i is the coefficient which indicated that the kernel principal component is spanned by feature space training samples. Multiplying $\bar{\Phi}(\mathbf{x}_j)$ with both sides of Eq. (2.38), the following equation is obtained

$$\lambda [\bar{\Phi}(\mathbf{x}_j) \mathbf{v}] = \bar{\Phi}(\mathbf{x}_j) \Sigma_{\mathbf{F}} \mathbf{v} \tag{2.40}$$

The problem can be transformed to Eq. (2.41) by introducing a kernel matrix $\bar{\mathbf{K}}_{ij} = \bar{\Phi}(\mathbf{x}_i) \cdot \bar{\Phi}^T(\mathbf{x}_j)$, where the kernel matrix is centered, α should be scaled as $\|\alpha\|^2 = 1/n\lambda$ to ensure the normality of $\mathbf{v}$ (Choi et al. 2005)

$$\lambda \alpha = \frac{1}{n} \bar{\mathbf{K}} \alpha \tag{2.41}$$

The score vector of qth observation is calculated as

$$\mathbf{t}_{q,k} = [\mathbf{v}_k \cdot \bar{\Phi}(\mathbf{x}_q)] = \sum_{i=1}^{n} \alpha_i^k [\bar{\Phi}(\mathbf{x}_q) \bar{\Phi}^T(\mathbf{x}_i)] = \sum_{i=1}^{n} \alpha_i^k \bar{\mathbf{K}}_{qi} \tag{2.42}$$

Then, T^2 and SPE statistics can be developed for process monitoring.

2.6 Conclusion

In this chapter, several typical multivariate statistical process control methods have been introduced, including PCA, PLS, ICA, FA, and KPCA. Based on these basic MSPC methods, more specific and complex process monioring methods can be further developed, some of which will be introduced in the following chapters.

Chapter 3
Non-Gaussian Process Monitoring

3.1 Introduction

Many successful applications have shown the practicability of multivariable statistical process control (MSPC). However, the achievable performance of the MSPC method is limited due to the assumption that monitored variables are Gaussian distribution. In fact, most of the process variables do not strictly form Gaussian distribution. In contrast, it is well known that many of the variables monitored in process systems are not independent; chemical processes are usually driven by fewer essential variables which may not be measured and the measured process variables may be the combinations of these independent latent variables. Independent component analysis (ICA) is an emerging technique for finding several independent variables as linear combinations of measured variables. What distinguished ICA from principal component analysis (PCA) is that it looks for components that are both statistically independent and non-Gaussian. PCA can only impose independence up to second-order statistics information, and the direction vectors are constrained to be orthogonal; whereas, ICA has no orthogonal constraint and also involves higher order statistics. ICA is able to extract the non-Gaussian information while PCA is looking for the Gaussian one. In other words, ICA may reveal more meaningful information in the non-Gaussian data than PCA. A number of applications of ICA have been reported in speech processing, biomedical signal processing, machine vibration analysis, nuclear magnetic resonance spectroscopy, infrared optical source separation, radio communications, etc. (Hyvarinen and Oja 2000). Recently, ICA has been introduced for the process monitoring purpose (Li and Wang 2002; Kano et al. 2003). Kano et al. (2004a) developed a unified framework for MSPC, which combined PCA-based SPC and ICA-based SPC. The proposed combined MSPC could monitor both of the Gaussian and non-Gaussian information of the process, good performance was shown in a multivariate system and a continuous stirred-tank reactor (CSTR) process. Several applications of the ICA method for process monitoring have also been reported (Lee et al. 2004a, b; Lee et al. 2006a; Zhang and Qin 2007; Ge and Song 2007, 2008; Zhao et al. 2008a; Hsu et al. 2010; Zhang and Zhang 2010; Odiowe and Cao 2010; Stefatos and Ben Hamza 2010).

Z. Ge, Z. Song, *Multivariate Statistical Process Control,* Advances in Industrial Control,
DOI 10.1007/978-1-4471-4513-4_3,

It is also important to note that all the reported ICA-based monitoring methods are based on the assumption that the essential variables which drive the process are non-Gaussian and discarded Gaussian parts are utilized to set up a single squared prediction error (*SPE*) monitoring chart. Actually, complex multivariate processes are always driven by non-Gaussian and Gaussian essential variables simultaneously. Separating the Gaussian information from the un-modelled Gaussian uncertainty will facilitate the diagnosis of faults which occur in different sources. Therefore, a two-step ICA-PCA information extraction strategy has been proposed and used for process monitoring.

Due to complicated confidence limit determination of the traditional ICA-based monitoring method, which is based on the kernel density estimation (KDE), support vector data description (SVDD) has been introduced to improve the determination of confidence limits for the non-Gaussian components (Tax and Duin 1999, 2004). Compared to KDE (Lee et al. 2004), the SVDD-based method is more computationally efficient, which relies on a quadratic programming cost function. To monitor discarded Gaussian parts of the process information, PCA might be the straightforward but not the optimal approach in which the projected directions have the maximum variations irrespective of the underlying information generating structure. More precisely, the process recordings always consist of inevitable measurement noises, while PCA is not capable of characterizing the different influences from common process Gaussian impelling factors and random variations due to measurement equipments. A probabilistic generative model (probabilistic PCA, PPCA) was employed to address this issue via the expectation and maximization (EM) algorithm (Kim and Lee 2003). However, PPCA assumes the same noise level of all measurement variables, which cannot be easily satisfied in practice. To this end, a general extension of PPCA model, factor analysis (FA) can model different noise levels of measurement variables. To determine the number of essential variables, lots of useful approaches have been reported, including Akaike Information Criterion (AIC), Bayesian Information Criterion (BIC), SCREE test, PRESS cross validation test (Chiang et al. 2001) and more recent fault detection criteria (Valle et al. 1999; Wang et al. 2002, 2004).

3.2 Two-step ICA-PCA Information Extraction Strategy for Process Monitoring

3.2.1 Process Monitoring Method

Giving the data matrix $\mathbf{X}$, assume r independent components (ICs) are extracted, $\mathbf{S} = [s_1, s_2, \ldots, s_r] \in \mathbf{R}^{r \times n}$. In order to monitor the non-Gaussian part of the process, a new statistic variable was defined (Lee et al. 2004; Ge and Song 2007)

$$I^2 = \mathbf{s}^T \mathbf{s} \tag{3.1}$$

After the non-Gaussian information has been extracted, the residual matrix $\boldsymbol{E}$ is obtained. Then PCA is used to analyse it, expanding $\boldsymbol{E}$ as follows

$$\mathbf{E} = \sum_{i=1}^{k} \mathbf{t}_i \mathbf{p}_i^T + \mathbf{F} \tag{3.2}$$

where $\mathbf{F}$ is the residual resulting from the PCA model. Here, we define the limits of T^2 and SPE statistics to monitor the remaining Gaussian part of the process

$$T^2 = \sum_{i=1}^{k} \frac{\mathbf{t}_i^2}{\lambda_i} \leq \frac{k(n-1)}{n-k} F_{k,(n-k),\alpha}, \tag{3.3}$$

$$SPE = \mathbf{f}\mathbf{f}^T = \mathbf{e}(\mathbf{I} - \mathbf{P}\mathbf{P}^T)\mathbf{e}^\mathrm{T} \leq SPE_{\mathrm{lim}} \tag{3.4}$$

In the PCA monitoring method, the confidence limits are based on a specified distribution, based upon the assumption that the latent variables follow a Gaussian distribution. However, in ICA monitoring, the independent component do not conform to a specific distribution, hence, the confidence limit of the I^2 statistic cannot be determined directly from a particular approximate distribution. An alternative approach to define the nominal operating region of the I^2 statistic is to use KDE (Chen et al. 2000, 2004). Here we only need a univariate kernel estimator, which is defined by

$$\hat{f}(I^2, \mathbf{H}) = \frac{1}{n} \sum_{i=1}^{n} K\left[\mathbf{H}^{-1/2}(I^2 - I_i^2)\right] \tag{3.5}$$

where $\mathbf{H}$ is the bandwidth matrix and K is a kernel function, which satisfies the following condition

$$K(I^2) \geq 0, \int_{R^p} K(I^2)dI^2 = 1 \tag{3.6}$$

There are a number of kernel functions, among which the Gaussian kernel function is the most commonly used.

Many methods have been proposed for the estimation of the window width or the smoothing parameter, which is of crucial importance in density estimation. One efficient method is called mean integrated squared error (MISE) (Hyvarinen and Oja 2000). There are three choices for $\mathbf{H}$

1. For a process with p variables, a full symmetrical bandwidth matrix

$$\mathbf{H} = \begin{bmatrix} h_{11}^2 & h_{12}^2 & \cdots & h_{1p}^2 \\ h_{21}^2 & h_{22}^2 & \cdots & h_{2p}^2 \\ \vdots & \vdots & & \vdots \\ h_{p1}^2 & h_{p2}^2 & \cdots & h_{pp}^2 \end{bmatrix}$$

which is a positive-definite matrix with $p(p+1)/2$ parameters, in which $h_{ik}^2 = h_{ki}^2$;

2. A diagonal matrix with only p parameters, $\mathbf{H} = \mathrm{diag}(h_1^2, h_2^2, \ldots, h_p^2)$;
3. A diagonal matrix with one parameter, $\mathbf{H} = h^2\mathbf{I}$, where I is an identity matrix.

Approach one is the most accurate but is unrealistic in terms of computational load and time. Approach three is the simplest, and the method used in the case of univariate data, and can be adopted with slight modification. However, it can cause problems in some situations due to the loss of accuracy by forcing the bandwidths in all dimensions to be the same. This can introduce significant error in the density function shape. Hence, approach two appears to be the appropriate choice, as a compromise, but the computational load is still unacceptable if the dimensionality of the problem is high and the size of the sample is large.

The implementation of the monitoring method, as mentioned in the previous section, consists of two procedures: off-line modelling and on-line monitoring. In the off-line modelling procedure, an ICA-PCA monitoring model is developed in the normal operating condition. Then, the fault detection is executed by using this monitoring model in the on-line monitoring procedure. The detailed algorithm flow of the monitoring procedure is summarized as follows

(1) *Off-line modelling procedure*

Step 1 Acquire an operating dataset $\mathbf{X}$ during normal process;
Step 2 Normalize and whiten the data matrix $\mathbf{X}$;
Step 3 Carry out the ICA algorithm to obtain the matrix $\mathbf{W}$, $\mathbf{B}$ and $\hat{\mathbf{S}}$, so that $\hat{\mathbf{S}}$ has great non-Gaussianity. We can also obtain $\mathbf{A} = (\mathbf{Q}^T\mathbf{Q})^{-1}\mathbf{Q}^T\mathbf{B}$ and $\mathbf{W} = \mathbf{B}^T\mathbf{Q}$;
Step 4 For each sample, calculate its I^2 value, $I^2(k) = \hat{\mathbf{s}}(k)^T\hat{\mathbf{s}}(k)$, $(k = 1, 2, \ldots, n)$, using KDE to decide its 99 or 95 % confidence limit $I_{\lim}^2$;
Step 5 Carry out the PCA algorithm to obtain the score matrix $\mathbf{T}$ and the load matrix $\mathbf{P}$, decide the confidence limits $T_{\lim}^2$ and $SPE_{\lim}$ for T^2 and SPE.

(2) *On-line monitoring procedure*

Step 1 For a new sample data $\mathbf{x}_{new}$, using the same scaling method used in the modelling steps;
Step 2 Calculate the ICs of the new sample data, $\mathbf{x}_{new}$, $\hat{\mathbf{s}}_{new} = \mathbf{W}\mathbf{x}_{new}$;
Step 3 Calculate the I^2 statistic value for this new sample data, $I_{new}^2 = \hat{\mathrm{S}}_{new}^T\hat{\mathrm{S}}_{new}$;
Step 4 The remaining part is given by $\mathbf{e}_{new} = \mathbf{x}_{new} - \mathbf{A}\hat{\mathbf{s}}_{new}$, calculate the score vector, $\mathbf{t}_{new} = \mathbf{e}_{new}\mathbf{P}$, and also obtain the residual, $\mathbf{f}_{new} = \mathbf{e}_{new} - \hat{\mathbf{e}}_{new}$;
Step 5 Calculate T_{new}^2 and SPE_{new} using Eqs. (3.3) and (3.4);
Step 6 If the statistics reject their corresponding limits, some kind of fault is detected, otherwise, go to step 1 and continue monitoring.

The on-line process fault detection scheme of the ICA-PCA based method is shown in Fig. 3.1.

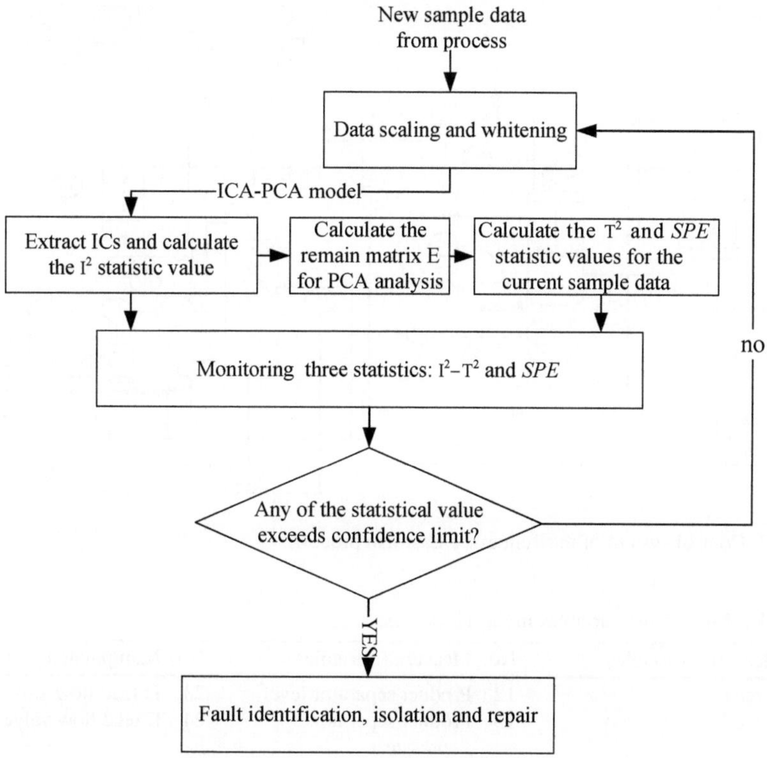

Fig. 3.1 Process monitoring strategy for the ICA-PCA method

3.2.2 *TE Benchmark Case Study*

As a benchmark simulation, the Tennessee Eastman (TE) process has been widely used to test the performance of various monitoring approaches (Downs and Vogel 1993; Chiang et al. 2001; Raich and Cinar 1995; Singhai and Seborg 2006). This process consists of five major unit operations: a reactor, a condenser, a compressor, a separator and a stripper. The control structure is shown schematically in Fig. 3.2 which is the second structure listed in Lyman and Georgakist (1995). The TE process has 41 measured variables (22 continuous process measurements and 19 composition measurements) and 12 manipulated variables, a set of 21 programmed faults are introduced to the process. The details on the process description are well explained in a book of Chiang et al. (2001).

In this process, the variables which are selected for monitoring are listed in Table 3.1. There are 33 variables in the table, 22 continuous process measurements and 11 manipulated variables. The agitation speed is not included because it is not manipulated. Besides, we also exclude 19 composition measurements as they are difficult to measure on-line in real processes. The simulation data which we have collected were separated into two parts: the training datasets and the testing datasets;

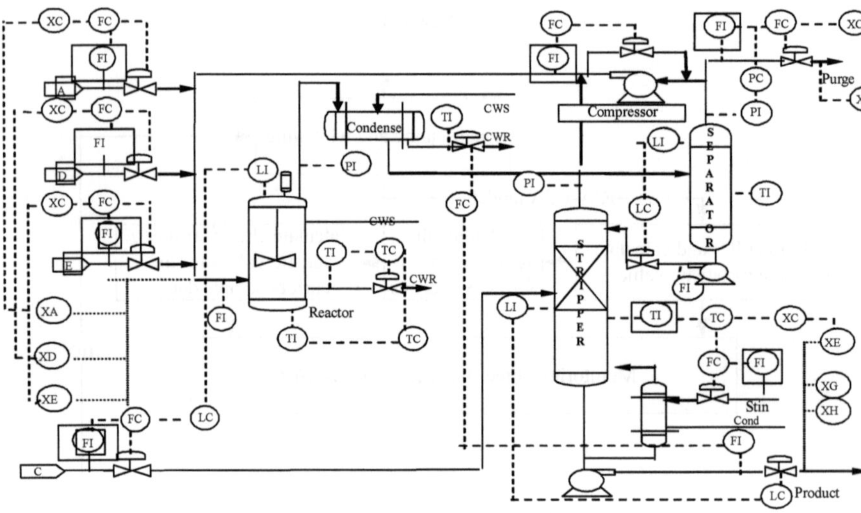

Fig. 3.2 Control system of the Tennessee Eastman process

Table 3.1 Monitoring variables in the TE process

No.	Measured variables	No.	Measured variables	No.	Manipulated variables
1	A feed	12	Product separator level	23	D feed flow valve
2	D feed	13	Product separator pressure	24	E feed flow valve
3	E feed	14	Product separator underflow	25	A feed flow valve
4	Total feed	15	Stripper level	26	Total feed flow valve
5	Recycle flow	16	Stripper pressure	27	Compressor recycle valve
6	Reactor feed rate	17	Stripper underflow	28	Purge valve
7	Reactor pressure	18	Stripper temperature	29	Separator pot liquid flow valve
8	Reactor level	19	Stripper steam flow	30	Stripper liquid product flow valve
9	Reactor temperature	20	Compressor work	31	Stripper steam valve
10	Purge rate	21	Reactor cooling water outlet temperature	32	Reactor cooling water flow
11	Product separator temperature	22	Separator cooling water outlet temperature	33	Condenser cooling water flow

they consisted of 960 observations for each mode (1 normal and 21 fault), respectively, and their sampling interval was 3 min. All faults in the testing datasets were introduced in the process at sample 160, which are tabulated in Table 3.2. Faults 1–7 are step changes of process variables; faults 8–12 are random changes of variables; fault 13 is slow shift of Reaction kinetics; faults 14, 15 and 21 are related to valve sticking and faults 16–20 are types of unknown faults. Among these faults, some faults are easy to detect as they greatly affect the process and change the relations

Table 3.2 Process disturbances

Fault number	Process variable	Type
1	A/C feed ratio, B composition constant (stream 4)	Step
2	B composition, A/C ratio constant (stream 4)	Step
3	D feed temperature (stream 2)	Step
4	Reactor cooling water inlet temperature	step
5	Condenser cooling water inlet temperature	Step
6	A feed loss (stream 1)	Step
7	C header pressure loss-reduced availability (stream 4)	Step
8	A, B, C feed composition (stream 4)	Random variation
9	D feed temperature (stream 2)	Random variation
10	C feed temperature (stream 4)	Random variation
11	Reactor cooling water inlet temperature	Random variation
12	Condenser cooling water inlet temperature	Random variation
13	Reaction kinetics	Slow drift
14	Reactor cooling water valve	Sticking
15	Condenser cooling water valve	Sticking
16	Unknown	Unknown
17	Unknown	Unknown
18	Unknown	Unknown
19	Unknown	Unknown
20	Unknown	Unknown
21	Valve position constant (stream 4)	Constant position

between process variables. However, there are also faults that are difficult to detect (faults 3, 9 and 15) because they are very small and have little influence to the process.

So far, 22 datasets have been generated, corresponding to the 22 different operation modes (1 normal and 21 fault) in the TE process. Before the application of PCA, ICA and ICA-PCA, all the datasets were auto-scaled. A total of 9 ICs and 15 principal components (PCs) were selected for ICA and PCA by cross-validation, respectively. In the study of the ICA-PCA method, we select the same number of ICs (9) and PCs (15) to compare the monitoring performance with ICA and PCA. The 99 % confidence limits of all the statistic variables were determined by KDE. For each statistic, the detection rates for all the 21 fault modes were calculated and tabulated in Table 3.3. The minimum missing detection rate achieved for each mode is marked with a bold number except the modes of faults 3 and 9; as these faults are quite small and have almost no effect on the overall process so they were excluded from our research. As shown in Table 3.3, ICA outperforms PCA for most fault modes, and particularly the missing detection rates of ICA for the modes of faults 5, 10 and 16 are much lower than that of PCA, which indicates that ICA can detect small events that are difficult to detect by PCA. However, the missing detection rates of ICA-PCA for most modes are even lower than that of ICA, as ICA-PCA not only monitors the non-Gaussian information of the process by ICA, but also the Gaussian information is monitored by PCA. As a result of the simulation, the best monitoring performance is found in the case of ICA-PCA. In most modes, the missing detection rate of ICA-PCA is the lowest.

The monitoring results of fault 5 are shown in Fig. 3.3. The condenser cooling water inlet temperature is step changed in the mode of fault 5. In that mode, the flow

Table 3.3 Monitoring results of the TE process

Fault modes	PCA T^2	PCA *SPE*	ICA I^2	ICA *SPE*	ICA-PCA I^2	ICA-PCA T^2	ICA-PCA *SPE*
1	0.007	0.004	0.006	0.002	0.005	0.003	*0*
2	0.022	0.015	0.020	0.010	0.018	0.015	*0.008*
3	1.000	0.998	0.995	0.990	0.996	0.993	0.985
4	0.975	0.045	0.035	*0*	0.032	0.020	*0*
5	0.750	0.745	*0*	*0*	*0*	*0*	*0*
6	0.010	*0*	*0*	*0*	*0*	*0*	*0*
7	0.090	*0*	0.065	*0*	0.050	0.035	*0*
8	0.035	0.025	0.030	0.021	0.028	0.022	*0.011*
9	1.000	0.997	0.995	0.996	0.994	0.993	0.991
10	0.710	0.651	0.215	0.200	0.212	*0.156*	0.162
11	0.811	0.367	0.756	*0.182*	0.712	0.526	*0.182*
12	0.031	0.027	0.025	*0*	0.024	0.005	*0*
13	0.065	0.050	0.025	*0*	0.016	0.011	*0*
14	0.162	0	0.045	*0*	0.042	0.013	*0*
15	0.999	0.985	0.426	0.328	0.412	0.268	*0.196*
16	0.812	0.755	0.213	*0.125*	0.185	0.146	*0.125*
17	0.241	0.112	0.211	0.196	0.206	*0.107*	0.195
18	0.112	0.100	0.102	0.085	0.100	*0.085*	0.098
19	0.998	0.762	0.351	0.322	0.346	0.238	*0.208*
20	0.712	0.524	0.426	0.358	0.412	*0.247*	0.306
21	0.725	0.621	0.635	0.528	0.622	0.549	*0.526*

rate of the outlet stream from the condenser to the separator also increases, which causes an increase in temperature in the separator, and thus affects the separator cooling water outlet temperature. As we have established a control structure for this process, the control loops will act to compensate for the change and the temperature in the separator will return to its set-point. It takes about 10 h to reach the steady state again. As shown in Fig. 3.3a, PCA has detected the fault at sample 160 approximately. However, the fault cannot be detected after approximately sample 350 as most of the variables returned to their set-points. In this case, if a process operator judges the status of the process based on PCA, they would probably conclude that a fault entered the process and then corrected itself in about 10 h. However, as a matter of fact, the condenser cooling water inlet temperature is still high than normal condition after sample 350, and the condenser cooling water flow rate is continuously manipulated while most variables return to their set-points. This indicates that a fault still remains in the process. As PCA is based on only second-order statistics information, the effect of the condenser cooling water flow may be neglected compared to the effect of other variables, thus, the problem occurs. Alternatively, as shown in Fig. 3.3b and 3.3c, all the ICA and ICA-PCA statistic variables stayed above their confidence limit, which indicate that a fault remains in the process. What distinguished ICA from PCA is that ICA involves higher order statistics information, which is non-Gaussian and, thus, can correctly reflect the effect of condenser cooling water flow rate. Such a persistent fault detection statistic will continue to inform the operator that a process abnormality remains in the process; although, all the process variables will appear to have returned to their normal values through control loops. In the

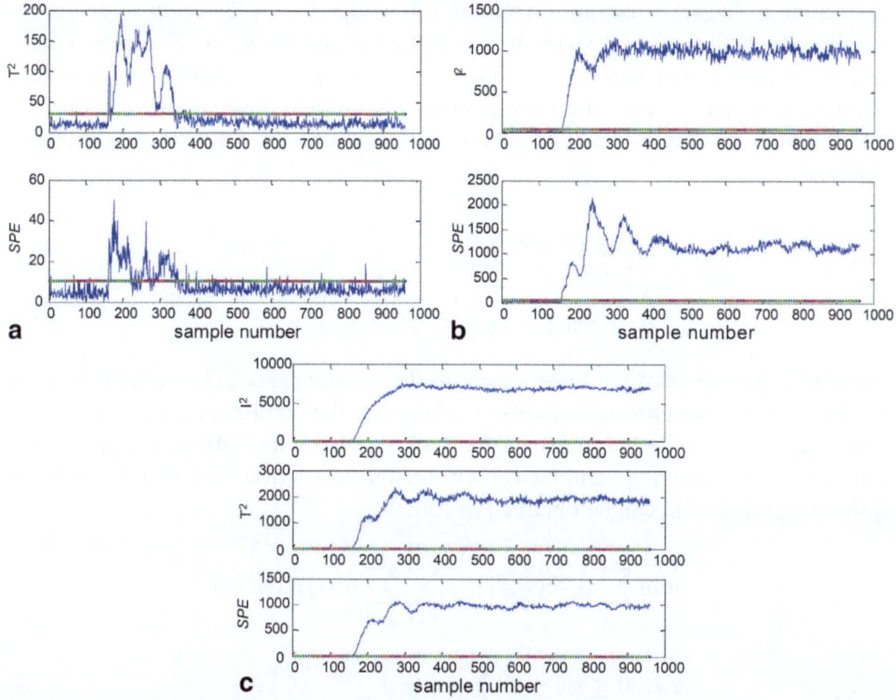

Fig. 3.3 Monitoring results for mode of fault 5 of TE process: **a** result based on PCA, **b** result based on ICA, **c** result based on ICA-PCA

method of ICA-PCA, as ICA has extracted essential components that underlie the process,the following procedure of PCA can reflect the effect of condenser cooling water flow rate. So, ICA and ICA-PCA may provide the process operator with more reliable information. Additionally, comparing the result of ICA-PCA with that of ICA, we can conclude that the effect of the fault is magnified more in the former statistic variables.

3.3 Process Monitoring Based on ICA-FA and SVDD

3.3.1 Support Vector Data Description

The main idea of SVDD is to map the input vectors to a feature space and to find hypersphere with the minimum volume which separates the transferred data from the rest of the feature space. Applications have shown a high generalization performance of SVDD if large reference dataset with few abnormal samples is available (Tax and Duin 1999).

Assume a dataset containing n samples $\hat{\mathbf{s}}_i \in \mathbf{R}^r, i = 1, 2, \ldots, n$ is given, the mapping from the original space to the feature space Φ: $\hat{\mathbf{s}} \rightarrow F$ can be simply done by a given kernel $K(\mathbf{s}_i, \mathbf{s}_j) = \langle \Phi(\mathbf{s}_i), \Phi(\hat{\mathbf{s}}_j) \rangle$, which computes the inner product in the feature space. Here, the most popular Gaussian kernel is used. To construct the minimum volume of the hypersphere, SVDD solves the following optimization problem

$$\min_{R,a,\xi} R^2 + C \sum_{i=1}^{n} \xi_i$$
$$s.t. \left\| \Phi(\hat{\mathbf{s}}_i) - \mathbf{a} \right\|^2 \le R^2 + \xi_i, \xi_i \ge 0 \tag{3.7}$$

where $\mathbf{a}$ is the center of the hypersphere in the feature space. The variable C gives the trade-off between the volume of the sphere and the number of errors (number of target objects rejected). ξ_i represents the slack variable which allows the probability that some of the training samples can be wrongly classified. The dual form of the optimization problem can be obtained as follows

$$\min_{\alpha_i} \sum_{i=1}^{n} \alpha_i K(\hat{\mathbf{s}}_i, \hat{\mathbf{s}}_j) - \sum_{i=1}^{n} \sum_{j=1}^{n} \alpha_i \alpha_j K(\hat{\mathbf{s}}_i, \hat{\mathbf{s}}_j)$$
$$s.t. \ 0 \le \alpha_i \le C, \sum_{i=1}^{n} \alpha_i = 1 \tag{3.8}$$

where α_i is Lagrange multipliers. After the hpyersphere in the feature space has been constructed, the hypothesis that a new sample $\hat{s}_{new}$ is normal is accepted if the distance $d[\Phi(\hat{\mathbf{s}}_{new})] = \left\| \Phi(\hat{\mathbf{s}}_{new}) - \mathbf{a} \right\| \le R$, which is given by

$$d[\Phi(\hat{\mathbf{s}}_{new})] = \sqrt{K(\hat{\mathbf{s}}_{new}, \hat{\mathbf{s}}_{new}) - 2\sum_{i=1}^{n} \alpha_i K(\hat{\mathbf{s}}_i, \hat{\mathbf{s}}_{new}) + \sum_{i=1}^{n} \sum_{j=1}^{n} \alpha_i \alpha_j K(\hat{\mathbf{s}}_i, \hat{\mathbf{s}}_j)} \tag{3.9}$$

3.3.2 Process Monitoring Based on ICA-FA and SVDD

As demonstrated, conventional statistical methods are under the assumption that processes are driven by fewer essential variables. These essential variables are either non-Gaussian or Gaussian distribution. However, under more general circumstances, processes are driven by non-Gaussian and Gaussian essential variables simultaneously, which formulates the key assumption of the mixed essential component analysis (ICA-FA) method. Figure 3.4 shows the difference between ICA-FA and the conventional methods (ICA and FA). The relationship of measurement variables and mixed essential variables is given as follows (Ge et al. 2009a)

$$\mathbf{x} = \mathbf{x}_1 + \mathbf{x}_2 = \mathbf{A}\mathbf{s} + \mathbf{P}\mathbf{t} + \mathbf{e} \tag{3.10}$$

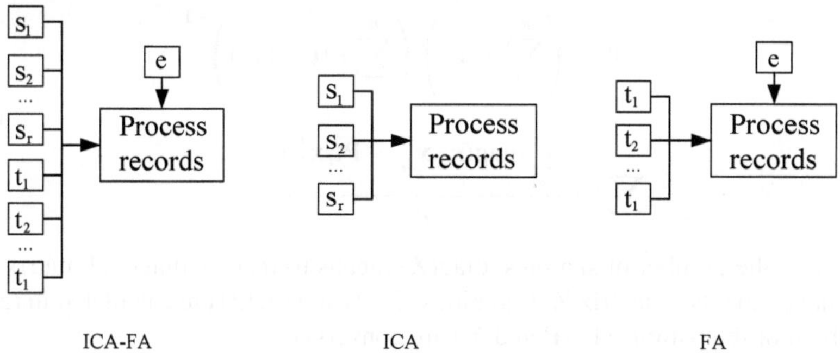

Fig. 3.4 Description of ICA-FA versus ICA and FA

where $\mathbf{x}_1$ and $\mathbf{x}_2$ represent non-Gaussian and Gaussian parts of the process, respectively, $\mathbf{A}$ is the mixing matrix of non-Gaussian essential variable $\mathbf{s}$, $\mathbf{P}$ is the loading matrix of Gaussian essential variables $\mathbf{t}$ and $\mathbf{e}$ is the Gaussian un-modelled uncertainty, relating to measurement noises, etc., with zero means and different variances $\Sigma = diag\{\lambda_i\}_{1,2,\ldots,m}$, which permits measurement variables to have different noise levels. As a general form of PPCA, FA performs better than PCA and PPCA. Since process measurements are always corrupted by noise, the ignorance of such information generation structure will make PCA method leave some essential information in the discarded space and trigger monitoring errors. In contrast, the noise structure is considered in PPCA and FA; thus, the useful information can be correctly extracted and monitored separately.

To estimate the new parameter set $\Theta = \{\mathbf{A}, \mathbf{P}, \Sigma\}$ and extract essential variables $\mathbf{EA} = \{\mathbf{s}, \mathbf{t}\}$ from measurement variables $\mathbf{x}$, a two-step estimation and extraction strategy can be developed. First, non-Gaussian essential variables can be extracted by the efficient FastICA or PSO-based method (Xie and Wu 2006), thus the mixing matrix $\mathbf{A}$ is obtained. Second, the parameter $\mathbf{P}$ and Σ can be estimated by the EM algorithm, which is described as follows. First, the mean and variance of the Gaussian part of measurement variables $\mathbf{x}_2$ can be calculated as

$$\begin{aligned} \mu^{x_2} &= \mathbf{P}\mathrm{E}[\mathbf{t}] + \mathrm{E}[\mathbf{e}] = 0; \\ \Sigma^{x_2} &= \mathbf{P}\mathrm{E}[\mathbf{t}\mathbf{t}^T]\mathbf{P}^T + \mathrm{E}[\mathbf{e}\mathbf{e}^T] = \mathbf{P}\mathbf{P}^T + \Sigma \end{aligned} \tag{3.11}$$

Hence, the distribution of $\mathbf{x}_2$ is $p(\mathbf{x}_2) = \mathrm{N}\{\mathbf{0}, \mathbf{P}\mathbf{P}^T + \Sigma\}$. EM is an iterative algorithm, it consists of two steps: E-step and M-step.

In E-step, two statistics of the Gaussian essential variable $\mathbf{t}$ are estimated

$$\begin{aligned} \hat{\mathbf{t}}_i &= \mathrm{E}[\mathbf{t}_i|\mathbf{x}_{2_i}] = \mathbf{Q}\mathbf{x}_{2_i} \\ \mathrm{E}[\mathbf{t}_i\mathbf{t}_i^T|\mathbf{x}_{2_i}] &= \mathbf{I} - \mathbf{Q}\mathbf{P} + \mathbf{Q}\mathbf{x}_{2_i}\mathbf{x}_{2_i}^T\mathbf{Q}^T \end{aligned} \tag{3.12}$$

where $\mathbf{Q} = \mathbf{P}^T(\mathbf{P}\mathbf{P}^T + \Sigma)^{-1}$.

In M-step, the parameters $\mathbf{P}$ and Σ are estimated so that the value of the likelihood function is maximized, the estimated parameters are calculated as

$$
\begin{aligned}
\mathbf{P} &= \left(\sum_{i=1}^{n} \mathbf{x}_{2_i} \hat{\mathbf{t}}_i^T\right) \left(\sum_{i=1}^{n} \mathrm{E}[\mathbf{t}_i \mathbf{t}_i^T | \mathbf{x}_{2_i}]\right)^{-1} \\
\sum &= \frac{\sum_{i=1}^{n} \mathrm{diag}(\mathbf{x}_{2_i} \mathbf{x}_{2_i}^T - \mathbf{P} \hat{\mathbf{t}}_i \mathbf{x}_{2_i}^T)}{n}
\end{aligned}
\tag{3.13}
$$

where n is the number of samples, diag($\mathbf{Z}$) means to make a diagonal matrix using diagonal elements of matrix $\mathbf{Z}$. Equations (3.12) and (3.13) are calculated iteratively until both of the parameters ($\mathbf{P}$ and Σ) are converged.

In summary, when the reference dataset $\mathbf{X} = \{\mathbf{x}_i\}_{i=1,2,\ldots,n}$ is available, with the ICA-FA model defined by Eq. (3.10), process information can be separated into three different parts. The systematic information includes non-Gaussian and Gaussian parts which are driven by their corresponding essential variables, whereas the noise information and other unmeasured disturbances are represented by $\mathbf{e}$.

As described in the previous section, the entire information of the process is partitioned into three parts: non-Gaussian systematic information, Gaussian systematic information and noise information. Therefore, three different monitoring statistics can be developed for fault detection. First, for monitoring non-Gaussian information, SVDD-based method is used. In contrast to the confidence limit determination method KDE, the SVDD reduces the complexity to quadratic programming problem that produces a unique analytical solution. Given the training dataset $\mathbf{S}$ (the extracted independent components), the hypersphere is constructed. The center $\mathbf{a}$ and the radius R can be determined by (Tax and Duin 2004)

$$
\begin{aligned}
\mathbf{a} &= \sum_{i=1}^{n} \alpha_i \Phi(\hat{\mathbf{s}}_i) \\
R &= \sqrt{1 - 2\sum_{i=1}^{n} \alpha_i K(\hat{\mathbf{s}}_z, \hat{\mathbf{s}}_j) + \sum_{i=1}^{n} \sum_{j=1}^{n} \alpha_i \alpha_j K(\hat{\mathbf{s}}_i, \hat{\mathbf{s}}_j)}
\end{aligned}
\tag{3.14}
$$

The non-Gaussian statistic (NGS) can be developed as the distance between the data sample and the center of the hypersphere, thus

$$
NGS_i = d^2[\Phi(\hat{\mathbf{s}}_i)] = \left\|\Phi(\hat{\mathbf{s}}_i) - \mathbf{a}\right\|^2 \le NGS_{\mathrm{lim}} = R^2 \tag{3.15}
$$

where $\hat{\mathbf{s}}_i$ is the extracted independent component of the new sample, $d[\Phi(\hat{\mathbf{s}}_i)]$ is given in Eq. (3.9). If $NGS_i \le R^2$, the non-Gaussian information is regarded as normal, else, it will be regarded as an outlier or some fault and disturbance has happened.

Since the distribution of Gaussian essential variables $\mathbf{t}$ is assumed to be N$\{\mathbf{0}, \mathbf{I}\}$, the Mahalanobis norm of $\mathbf{t}$ is identical to the Euclidian norm. Different from the projection method, the Gaussian essential variables cannot be calculated directly. However, it can be substituted by their estimation, which is given in Eq. (3.12).

It is known that the squared Mahalanobis norm of $\mathbf{t}$ follows χ^2 distribution, the confidence limit of T^2 statistic can be determined as follows

$$T_i^2 = \left\|\hat{\mathbf{t}}_i\right\|^2 = \mathbf{x}_{2_i}^T \mathbf{Q}^T \mathbf{Q} \mathbf{x}_{2_i} \leq \chi_{\alpha,l}^2 \tag{3.16}$$

where $\chi_{\alpha,l}^2$ represent the confidence limit with α confidence and l is the corresponding free parameter. If the value of T^2 statistic goes beyond the confidence limit, the Gaussian systematic process is regarded out-of-control.

Similarly, the noise part of the process information can also be monitored by the squared Mahalanobis distance. Hence, the Gaussian information matrix $\mathbf{x}_2$ can be represented as

$$\mathbf{x}_2 = \mathbf{Pt} + \mathbf{e} \tag{3.17}$$

Notice Eqs. (3.12) and (3.17), the noise information $\mathbf{e}$ can be estimated as follows

$$\hat{\mathbf{e}}_i = \mathrm{E}[\mathbf{e}_i | \mathbf{x}_{2_i}] = (\mathbf{I} - \mathbf{PQ})\mathbf{x}_{2_i} \tag{3.18}$$

Then, the confidence limit of *SPE* statistic can be determined by

$$SPE_i = \left\|\Sigma^{-1/2}\mathbf{e}_i\right\|^2 = \mathbf{x}_{2_i}^T (\mathbf{I} - \mathbf{PQ})^T \Sigma^{-1} (\mathbf{I} - \mathbf{PQ})\mathbf{x}_{2_i} \leq \chi_{\alpha,m}^2 \tag{3.19}$$

where $\chi_{\alpha,m}^2$ represent the confidence limit with α confidence and m is the corresponding free parameter. If the value of *SPE* statistic goes beyond the confidence limit, the noise part of the process is regarded abnormal. This may be due to the change in noise level, some unmeasured disturbances and other process structure changes.

Note that these two Gaussian parts of information both follow the χ^2 distribution; they can be monitored together. According to Eq. (3.11), the entire Gaussian statistic (*GS*) can be constructed as follows

$$GS_i = \left\|(\mathbf{PP}^T + \Sigma)^{-1/2}\mathbf{x}_{2_i}\right\|^2 = \mathbf{x}_{2_i}^T (\mathbf{PP}^T + \Sigma)^{-1} \mathbf{x}_{2_i} \leq \chi_{\alpha,m}^2 \tag{3.20}$$

Therefore, if changes happen in Gaussian part, the entire monitored *GS* should go beyond its confidence limit. However, if one wants to isolate the change between the systematic Gaussian part and the noise part, T^2 and *SPE* statistics should be chosen.

In summary, four statistics (*NGS*, T^2, *SPE* and *GS*) have been developed for process monitoring. By monitoring *NGS* and *GS*, the change between non-Gaussian and Gaussian information can be isolated. Furthermore, the change of Gaussian information can be isolated by monitoring T^2 and *SPE* separately. Hence, three types of changes can be monitored separately by their corresponding statistics. Change of any type or types of information will be reflected in related monitoring charts.

3.3.3 Case Study

In this section, the performance of the method is tested through the TE process. The variables selected for monitoring are the same as those used for control structure

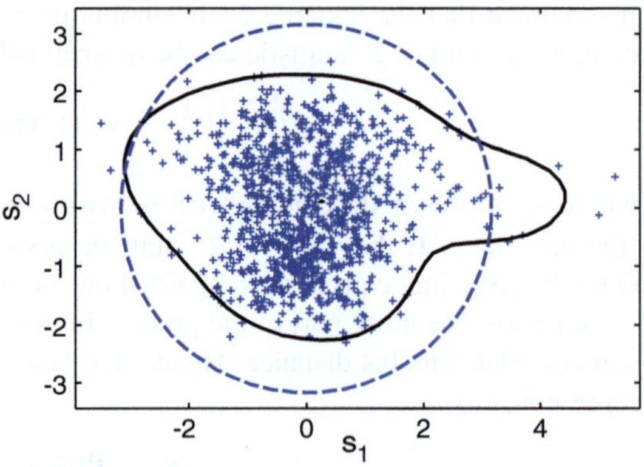

Fig. 3.5 Scatter plot of TE normal operation condition

development in Lyman and Georgakist (1995). Simulation data which are acquired in the normal process consist of 960 samples, and their sampling interval was 3 min. In order to evaluate fault detection and identification performance of the method, simulation data under 21 fault processes are also obtained. During the simulation, each fault is introduced in the 161st sample.

For monitoring of the TE process, the number of ICs and principal factor (PFs) should be chosen appropriately. In this study, 3 ICs and 6 PFs are selected. Then the ICA-FA model can be developed. The kernel parameter for the Gaussian function is chosen as $\sigma = 2.0$, the variable C is selected as $C = 0.0625$. For comparison, the ICA, PCA and ICA-PCA algorithms are also carried out. The number of ICs for ICA is selected as 3, whereas the PCs number for PCA is chosen as 6. All confidence limits are set as 99 %. The scatter plot of the first independent component versus the second independent component and two confidence limits of NGS and I^2 is given in Fig. 3.5. Under normal operation, both of the confidence limits indicate that 99 % samples are inside their corresponding regions. However, as discussed previously, the confidence limit of NGS is tighter than that of I^2, which is shown correctly in the figure. Therefore, when some fault happens, NGS is more sensitive to detect this fault. As samples lie in the gap between the NGS confidence limit and the I^2 confidence limit, only NGS can detect them. The I^2 statistic regards it as a normal operation because the confidence limit has not been rejected.

For each statistic, the monitoring results (type II error) are tabulated in Table 3.4. The minimum value achieved for each mode is marked with a bold number. As shown in the table, ICA-FA outperforms ICA, PCA and ICA-PCA for most of the fault modes. Particularly, the missing detection rates of ICA-FA for faults 10, 15, 16 and 19–21 are much lower than those of the other three methods. As indicated in the 2–4 columns of the table, best monitoring performance of some faults are revealed by the NGS, some are gained by T^2, whereas others are gained by the SPE statistic, which shows the fault isolation ability of the ICA-FA method.

Table 3.4 Monitoring results of TE process (type II error)

Mode	ICA-FA			ICA		PCA		ICA-PCA		
	NGS	T^2	*SPE*	I^2	*SPE*	T^2	*SPE*	I^2	T^2	*SPE*
0	0.990	0.994	0.994	0.988	0.991	0.986	0.988	0.988	0.991	0.989
1	0.015	*0.003*	0.005	*0.003*	0.005	0.008	0.005	*0.003*	0.011	*0.003*
2	0.018	0.018	0.021	0.023	0.018	0.034	0.015	0.023	0.018	*0.016*
3	*0.911*	0.993	0.995	0.974	0.983	0.984	0.974	0.974	0.990	0.979
4	*0.963*	0.986	0.994	0.990	0.989	0.990	0.981	0.990	0.993	0.984
5	*0.764*	0.778	0.855	0.790	0.773	0.773	0.814	0.790	0.785	0.868
6	0.008	*0.000*	0.003	0.014	*0.000*	0.001	*0.000*	0.014	0.007	*0.000*
7	*0.596*	0.664	0.811	0.641	0.670	0.618	0.735	0.641	0.681	0.751
8	0.050	*0.029*	0.118	0.089	0.031	0.059	0.039	0.089	0.064	0.069
9	*0.933*	0.985	0.995	0.981	0.976	0.980	0.974	0.981	0.988	0.975
10	0.385	0.336	*0.271*	0.579	0.596	0.685	0.728	0.579	0.841	0.519
11	0.886	0.789	0.556	0.961	0.605	0.859	0.579	0.961	0.815	*0.538*
12	0.124	*0.009*	0.155	0.081	0.024	0.029	0.088	0.081	0.059	0.100
13	0.060	0.074	0.083	0.068	0.074	0.064	*0.056*	0.068	0.098	0.079
14	0.061	0.001	*0.000*	0.100	0.001	0.058	*0.000*	0.100	0.001	*0.000*
15	*0.900*	0.978	0.993	0.976	0.975	0.968	0.975	0.976	0.984	0.980
16	*0.664*	0.853	0.841	0.748	0.886	0.788	0.850	0.748	0.913	0.929
17	0.261	0.103	*0.039*	0.243	0.059	0.184	0.046	0.243	0.179	0.041
18	*0.091*	0.101	0.106	0.105	0.104	0.108	0.101	0.105	0.109	0.098
19	0.869	0.981	*0.809*	0.971	0.814	0.905	0.851	0.971	0.829	0.866
20	*0.449*	0.490	0.454	0.594	0.594	0.780	0.684	0.594	0.818	0.524
21	*0.374*	0.806	0.755	0.434	0.853	0.689	0.554	0.434	0.921	0.874

3.4 Conclusions

In this chapter, different non-Gaussian process monitoring methods have been reviewed, especially the ICA-based methods. For processes which are simultaneously driven by non-Gaussian and Gaussian components, a two-step ICA-PCA information extraction strategy has been demonstrated, based on which a corresponding monitoring scheme has been developed. Compared to the traditional ICA and PCA monitoring methods, the monitoring performance has been improved by the ICA-PCA based approach. Furthermore, the SVDD method has been introduced for modelling the distribution of the independent components, based on which a non-Gaussian monitoring statistic has been developed. Compared to the conventional I^2 monitoring statistic, the monitoring performance has been improved by this non-Gaussian monitoring statistic. Besides, for probabilistic monitoring of the Gaussian information of the process, the FA method has been incorporated. Two additional Gaussian monitoring statistics have been constructed for process monitoring. Through the simulation study of the TE benchmark process, the efficiency of both ICA-PCA and ICA-FA with SVDD methods have been demonstrated.

Chapter 4
Fault Reconstruction and Identification

4.1 Introduction

Fault detection and diagnosis are of ever growing importance for guaranteeing a safe and efficient operation of complex industrial processes. Over the past few decades, statistical-based techniques have been intensely researched, as they directly address the challenges of large recorded variable sets resulting from the increasing complexity of distributed control systems. Notable algorithmic developments have been firmly made based on principal component analysis (PCA) and partial least squares (PLS; Qin 2003; Venkatasubramanian et al. 2003).

Whilst statistically based fault detection schemes are well-established, several fault diagnosis methods have also been proposed, such as contribution plot and the backward elimination sensor identification (BESI) algorithm (Westerhuis et al. 2000; Stork et al. 1997). However, these methods are difficult to implement in practice given that the contribution method may produce misleading results (Lieftucht et al. 2006) and the BESI algorithm requires a new PCA model to be calculated for each step (Stork et al. 1997). For enhanced fault isolation, Gertler et al. (1999) introduced a structure residual-based approach, which was later improved upon by Qin and Li (2001).

Dunia and Qin (1998a, b) proposed a uniform geometric method for both unidimensional and multidimensional fault identification and reconstruction. This projection-based variable reconstruction work relied on the conventional *SPE* statistic. More recent work by Wang et al. (2002) discussed the detectability of faults for the PCA method using the T^2 statistic and Yue and Qin (2001) suggested the use of a combined index for reconstruction-based fault identification and isolation method. Other extensions of these methods include Lieftucht et al. (2006) and Li and Rong (2006). However, these methods rely on PCA, which is applicable under the assumption that the recorded process variables follow a multivariate Gaussian distribution.

For monitoring processes that exhibit non-Gaussian behavior, recent work on independent component analysis (ICA) has shown its potential (Lee et al. 2004a, b; Ge and Song 2007; Liu et al. 2008). Although ICA has been successfully employed

Z. Ge, Z. Song, *Multivariate Statistical Process Control,* Advances in Industrial Control,
DOI 10.1007/978-1-4471-4513-4_4, © Springer-Verlag London 2013

to extract non-Gaussian source signal that can be used to construct univariate monitoring statistics for fault detection, the issue of fault identification and isolation has only been sporadically touched upon. A notable exception is discussed in Lee et al. (2004a), where contribution plots for fault identification were developed for ICA models. However, as mentioned above, contribution plot may produce misleading and erroneous fault diagnosis results.

This chapter introduces a technique that relies on PCA-based fault reconstruction to address the important issue of fault diagnosis for multivariate non-Gaussian processes (Ge and Song 2009b). This method was developed for a description of the independent components in the feature space determined by the support vector data description (SVDD). In a similar fashion to PCA-based reconstruction, this scheme relies on predefined fault directions, where the effect of a fault upon these direction is evaluated using a developed fault diagnosis index.

For fault identification, Zhang et al. (1996) proposed the characteristic direction method to identify faults, which took the first loading vector as the characteristic direction. Although the first principal component extracted most of the fault information, lot of other information has been omitted. Krzanowski (1979) developed a method for measuring the similarity of two datasets using a PCA similarity factor S_{PCA}. Johannesmeyer et al. (2002) used that PCA similarity factor to identify the similarity between different operation modes. Within that method, all the information that the PCA model had covered was used, so the power of fault identification has been improved. However, these similarity factors are all based on the assumption that the data formed Gaussian distribution. For non-Gaussian process data, an ICA similarity factor has been defined (Ge and Song 2007). Based on the ICA-FA method in Chap. 3 a further similarity factor for the noise subspace of the process data has also been defined. These similarity indices are used to classify various different process faults.

4.2 Fault Reconstruction

4.2.1 Non-Gaussisan Fault Reconstruction Based on SVDD

For the introduction of the fault reconstruction method, we assume that the fault subspace is known *a priori*. This is not a restriction of generality, as the application of a singular value decomposition can estimate this subspace (Qin 2003). Assuming there are a total of J possible fault conditions, denoted here by $\{\mathrm{F}_j, j = 1, 2, \ldots, J\}$, each fault condition is described by an individual fault subspace. These subspaces are of dimension $\{\dim(\boldsymbol{\Theta}_j) \geq 1, j = 1, 2, \ldots, J\}$, which implies that unidirectional, $\dim(\boldsymbol{\Theta}_j) = 1$, and multidimensional fault conditions $\dim(\boldsymbol{\Theta}_j) > 1$ can be reconstructed.

After defining the fault subspace, $\mathbf{z}^*$ can be reconstructed from the corrupted data vector $\mathbf{z}$ using the ICA model and the SVDD method. Suppose a fault F_j has occurred, a reconstructed $\mathbf{z}_j$ represents an adjustment of the corrupted value $\mathbf{z}$ moving along the corresponding fault direction $\boldsymbol{\Theta}_j$

$$\mathbf{z}_j = \mathbf{z} - \boldsymbol{\Theta}_j \mathbf{f}_j \tag{4.1}$$

where $\mathbf{f}_j$ is a vector storing the estimated fault magnitude such that $\mathbf{z}_j$ is the closest to the normal region. Traditionally, the optimal reconstruction is obtained by minimizing $\|\mathbf{z}_j - \mathbf{z}_j^*\|$. For this method, the reconstruction is realized by moving the corrupted value $\Phi(\hat{\mathbf{s}})$ along the defined fault direction $\boldsymbol{\Theta}_j$ into the SVDD feature space. The reconstruction technique therefore leads to the following problem

$$\mathbf{f}_j = \arg\min_{\mathbf{f}} ||\Phi(\hat{\mathbf{s}}^*) - \mathbf{a}|| = \arg\min_{\mathbf{f}} ||\Phi(\hat{\mathbf{s}} - \mathbf{W}\boldsymbol{\Theta}_j\mathbf{f}) - \mathbf{a}|| \quad \hat{\mathbf{s}} = \mathbf{W}\mathbf{z} \quad \hat{\mathbf{s}}^* = \mathbf{W}\mathbf{z}^* \tag{4.2}$$

However, the optimization problem of Eq. (4.2) is problematic, since the formulation of the mapping function $\Phi(\cdot)$ is unknown. To overcome this deficiency, the iterative denoising and reconstruction approach developed for kernel PCA (Mika et al. 1998; Takahashi and Kurita 2002) can be inserted into Eq. (4.2). With $N < n$ support vectors, obtained by the SVDD method, the length of $\Phi(\hat{\mathbf{s}}) - \mathbf{a}$ is (Tax and Duin 2004)

$$||\Phi(\hat{\mathbf{s}}) - \mathbf{a}|| = \sqrt{K(\hat{\mathbf{s}}, \hat{\mathbf{s}}) - 2\sum_{i=1}^{N} \alpha_i K(\mathbf{sv}_i, \hat{\mathbf{s}}) + \sum_{i=1}^{N}\sum_{j=1}^{N} \alpha_i\alpha_j K(\mathbf{sv}_i, \mathbf{sv}_j)} \tag{4.3}$$

where $\hat{\mathbf{sv}}_i$ is the ith support vector obtained by the SVDD method and α_i is the corresponding coefficient. As this work utilizes the Gaussian kernel function $K(\hat{\mathbf{s}}_i, \hat{\mathbf{s}}_j) = \exp(-||\hat{\mathbf{s}}_i - \hat{\mathbf{s}}_j||^2/\delta)$, $K(\hat{\mathbf{s}}, \hat{\mathbf{s}}) = 1$, and $\sum_{i=1}^{N}\sum_{j=1}^{N} \alpha_i\alpha_j K(\hat{\mathbf{sv}}_i, \hat{\mathbf{sv}}_j)$ is a constant. By incorporating Eq. (4.3), the optimization problem in Eq. (4.2) finally becomes

$$\mathbf{f}_j = \arg\max_{\mathbf{f}} \sum_{i=1}^{N} \alpha_i K(\mathbf{sv}_i, \mathbf{s}_j^*) = \arg\max_{\mathbf{f}} \sum_{i=1}^{N} \alpha_i K(\mathbf{sv}_i, \hat{\mathbf{s}} - \mathbf{W}\boldsymbol{\Theta}_j\mathbf{f}) \tag{4.4}$$

Denoting $\boldsymbol{\Omega} = \sum_{i=1}^{N} \alpha_i K(\mathbf{sv}_i, \mathbf{W}\boldsymbol{\Theta}_j\mathbf{f})$ and taking the first partial with respect to $\mathbf{f}$ yields

$$\nabla_{\mathbf{f}}\boldsymbol{\Omega} = \frac{\partial\boldsymbol{\Omega}}{\partial\mathbf{f}} = \mathbf{0} \tag{4.5}$$

which for Gaussian kernel functions is equal to

$$\nabla_{\mathbf{f}}\boldsymbol{\Omega} = [\mathbf{W}\boldsymbol{\Theta}_j]^T \sum_{i=1}^{N} \alpha_i K(\mathbf{sv}_i, \hat{\mathbf{s}} - \mathbf{W}\boldsymbol{\Theta}_j\mathbf{f})(\hat{\mathbf{s}} - \mathbf{W}\boldsymbol{\Theta}_j\mathbf{f} - \mathbf{sv}_i) = \mathbf{0} \tag{4.6}$$

which can be simplified to

$$\nabla_{\mathbf{f}}\boldsymbol{\Omega} = [\mathbf{W}\boldsymbol{\Theta}_j]^T\mathbf{W}\boldsymbol{\Theta}_j \sum_{i=1}^{N} \alpha_i K(\mathbf{sv}_i, \hat{\mathbf{s}} - \mathbf{W}\boldsymbol{\Theta}_j\mathbf{f})\mathbf{f} +$$

$$[\mathbf{W}\boldsymbol{\Theta}_j]^T \sum_{i=1}^{N} \alpha_i K(\mathbf{sv}_i, \hat{\mathbf{s}} - \mathbf{W}\boldsymbol{\Theta}_j\mathbf{f})(\mathbf{sv}_i - \hat{\mathbf{s}}) = 0. \tag{4.7}$$

By rewriting Eq. (4.7), the unknown fault vector **f** can be estimated as follows

$$\mathbf{f} = [\boldsymbol{\Theta}_j{}^T \mathbf{W}^T \mathbf{W} \boldsymbol{\Theta}_j]^{-1} \frac{\boldsymbol{\Theta}_j{}^T \mathbf{W}^T \sum_{i=1}^{N} \alpha_i K(\mathbf{sv}_i, \hat{\mathbf{s}} - \mathbf{W}\boldsymbol{\Theta}_j \mathbf{f})(\hat{\mathbf{s}} - \mathbf{sv}_i)}{\sum_{i=1}^{N} \alpha_i K(\mathbf{sv}_i, \hat{\mathbf{s}} - \mathbf{W}\boldsymbol{\Theta}_j \mathbf{f})} \tag{4.8}$$

Appendix shows that the estimated fault vector $\mathbf{f}_j$ can be found iteratively

$$\mathbf{f}(t+1) = [\boldsymbol{\Theta}_j{}^T \mathbf{W}^T \mathbf{W} \boldsymbol{\Theta}_j]^{-1} \frac{\boldsymbol{\Theta}_j{}^T \mathbf{W}^T \sum_{i=1}^{N} \alpha_i K(\mathbf{sv}_i, \hat{\mathbf{s}} - \mathbf{W}\boldsymbol{\Theta}_j \mathbf{f}(t))(\hat{\mathbf{s}} - \mathbf{sv}_i)}{\sum_{i=1}^{N} \alpha_i K(\mathbf{sv}_i, \hat{\mathbf{s}} - \mathbf{W}\boldsymbol{\Theta}_j \mathbf{f}(t))} \tag{4.9}$$

such that

$$\mathbf{z}_j^* = \mathbf{z} - \boldsymbol{\Theta}_j \mathbf{f}_j^* \tag{4.10}$$

4.2.2 *Fault Diagnosis*

It is important to assess the impact of reconstructing each fault condition $\{\mathrm{F}_j, j = 1, 2, \ldots, J\}$. After reconstructing each of the possible fault conditions, the correct fault direction will allow the reconstruction procedure to shift the reconstructed sample $\mathbf{z}_j^*$ closest to the center of the SVDD sphere in the feature space. This, in turn, implies that the value of the NSG_j reduces significantly and shows an in-statistical-control situation if F_j is the correct fault condition. In contrast, if an incorrect direction is applied $\{\mathrm{F}_i, i = 1, 2, \ldots, J,\ i \neq j\}$, the univariate NGS_i statistic still violates its control limit. Calculating the non-Gaussian statistic value NGS_{j} by using Eqs. (4.11) and (4.13)

$$\hat{\mathbf{s}}_j^* = \mathbf{W}\hat{\mathbf{z}}_j^* \tag{4.11}$$

$$\left\| \Phi(\hat{\mathbf{s}}_j^*) - \mathbf{a} \right\| = \sqrt{K(\hat{\mathbf{s}}_j^*, \hat{\mathbf{s}}_j^*) - 2\sum_{i=1}^{N} \alpha_i K(\mathbf{sv}_i, \hat{\mathbf{s}}_j^*) + \sum_{i=1}^{N}\sum_{j=1}^{N} \alpha_i \alpha_j K(\mathbf{sv}_i, \mathbf{sv}_j)} \tag{4.12}$$

$$NGS_j = \left\| \Phi(\hat{\mathbf{s}}_j^*) - \mathbf{a} \right\|_2^2 \tag{4.13}$$

the impact of the reconstruction procedure upon the jth fault direction therefore relies on ξ_j

$$\xi_j = \frac{NGS_j}{NGS_{\lim}} \tag{4.14}$$

which is referred to here as the fault diagnosis index and $NGS_{\lim}$ is the control limit of the *NGS* statistic. If the reconstruction procedure has been applied using the correct fault direction, Eq. (4.14) highlights that a fault diagnosis index produces a value that is smaller than or equal to 1. This, in turn, implies that the faulty sample has been shifted along the correct fault direction to fall within the normal region. However, if an incorrect fault direction is applied this value still exceeds 1.

4.2.3 Simulation Example

The six process variables are simulated as linear combinations of a total of three source variables. In order to account for measurement uncertainty, normally distributed measurement noise was superimposed onto this matrix vector product

$$\mathbf{z} = \mathbf{A}\mathbf{s} + \mathbf{e}. \tag{4.15}$$

The linear combinations are governed by the mixing matrix **A**, which was randomly selected to be

$$\mathbf{A} = \begin{bmatrix} 0.815 & 0.906 & 0.127 & 0.913 & 0.632 & 0.098 \\ 0.279 & 0.547 & 0.958 & 0.965 & 0.158 & 0.971 \\ 0.957 & 0.485 & 0.800 & 0.142 & 0.422 & 0.916 \end{bmatrix}^T. \tag{4.16}$$

The $\mathrm{s} \in \mathrm{R}^3$ is a vector of source signal each of which follows a uniform distribution with [0 1] and **e** are Gaussian distributed i.i.d. sequences of zero mean and variance 0.01, $\mathrm{e} \sim \mathrm{N}\ \{0, 0.01\mathbf{I}\}$. To test the performance of the fault reconstruction and diagnosis method, a total of 1,000 samples were simulated. The first 500 samples served as reference data, whilst the remaining 500 samples were used for model validation and to inject fault conditions for studying the performance of the fault reconstruction scheme (test dataset). Selecting the parameters of the Gaussian kernel functions of the SVDD as $C = 0.08$ and $\sigma = 6.5$ produced the control limit of the NSG statistic for a significance level of 5 %.

The first fault condition was a bias of magnitude 2 injected to the second sensor in the test dataset after 250th sample. The conventional ICA-SVDD (Liu et al. 2008) method could successfully detect this fault. Next, applying the fault identification and isolation method and ICA contribution plots produced the plots shown in Fig. 4.1. As discussed in the Sect. 4.2.2, values below or equal to 1 imply this fault condition is responsible for the detected abnormal event. Values above 1 indicate that this fault condition cannot be considered as a potential root cause. Given the way in which the fault was injected the fault identification index for reconstructing along the direction of variable 2 should be identified as significant, which plot a confirms. However, the identification results of the contribution plot (ICA) in Fig. 4.1, plots 4.1b and 4.1c, show that most variables show a significant response to this event for the $I^2{}_{cont,j}$ and variables 4 and 5 for $SPE_{cont,\,j}$.

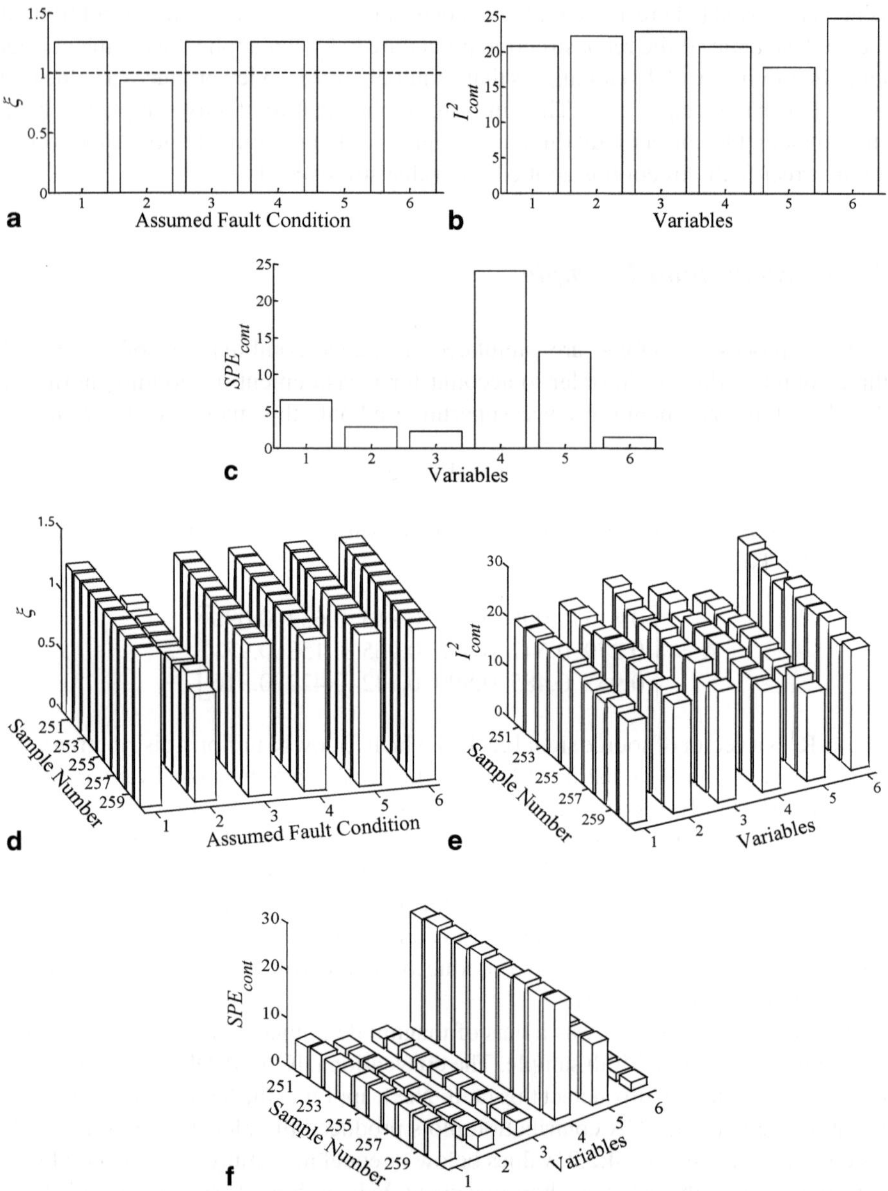

Fig. 4.1 Average results for 250 samples describing bias in sensor 2. **a** SVDD reconstruction. **b** ICA contribution plot. **c** SPE contribution plot and individual plots for first ten samples. **d** SVDD reconstruction. **e** ICA contribution plot. **f** SPE contribution plot

To analyze this misleading result in more detail, the way in which the variable contributions are generated need to be reexamined

$$I^2_{cont,j} \propto \mathbf{e}_j^T \mathbf{Q}^{\dagger} \mathbf{B}\hat{\mathbf{s}} = \mathbf{e}_j^T \mathbf{Q}^{\dagger} \mathbf{B}\mathbf{B}^T \mathbf{Q}\mathbf{z} = \mathbf{e}_j^T \boldsymbol{\Omega}\mathbf{z}. \tag{4.17}$$

If $m^* = m$ and $d = m$, it follows that $\boldsymbol{\Omega} = \mathbf{I}$ otherwise $\boldsymbol{\Omega}$ reduces to a matrix of rank $\min\{m^*, d\}$. The variable contributions to the SPE statistic are given by

$$SPE_{cont,j} = \left(\mathbf{e}_j^T [\mathbf{I} - \mathbf{A}\mathbf{B}^T\mathbf{Q}]\mathbf{z}\right)^2 = \left(\mathbf{e}_j^T \Pi \mathbf{z}\right)^2. \tag{4.18}$$

If we now define $\mathbf{z}$ by $\mathbf{z}_0 + \Delta\mathbf{z}$ where $\mathbf{z}_0$ is the measured vector without the impact of the fault condition and $\Delta\mathbf{z}$ is a vector that describes the fault impact, Eqs. (4.17) and (4.18) give rise to

$$\begin{aligned} I^2_{cont,j} &\propto \mathbf{e}_j^T \boldsymbol{\Omega}(\mathbf{z}_0 + \Delta\mathbf{z}) = I^2_{cont,j} + \Delta I^2_{cont,j} \\ SPE^2_{cont,j} &= (\mathbf{e}_j^T \Pi(\mathbf{z}_0 + \Delta\mathbf{z}))^2 \\ &= SPE^2_{cont,j} + \Delta SPE_{0cont,j} + \sqrt[2]{SPE_{0_{cont,j}} \Delta SPE_{cont,j}}. \end{aligned} \tag{4.19}$$

This separation shows that the fault contributions to the ICA-I^2 and -SPE statistics depend on $\mathbf{e}_j^T \boldsymbol{\Omega}\Delta\mathbf{z}$ and $\mathbf{e}_j^T \boldsymbol{\Pi}\Delta\mathbf{z}$, respectively. Assuming that the ith sensor is faulty, the contribution of this sensor bias upon the residual variables e_j is as follows

$$\Delta I^2_{cont,j} = \boldsymbol{\omega}_j^T \Delta\mathbf{z} = \sum_{i=1}^{m} \omega_{ij}\Delta z_i \;\; \Delta SPE_{cont,j} = \boldsymbol{\gamma}_j^T \Delta\mathbf{z} = \sum_{i=1}^{m} \boldsymbol{\gamma}_{ij}\Delta z_i \tag{4.20}$$

where Δz_i is the magnitude of the sensor bias, $\boldsymbol{\omega}_j^T$ and $\boldsymbol{\gamma}_j^T$ is the jth row vector of $\boldsymbol{\Omega}$ and $\boldsymbol{\Pi}$ whose ith elements are $\boldsymbol{\Omega}_{ij}$ and $\boldsymbol{\gamma}_{ij}$, respectively. As no assumptions can generally be imposed on $\boldsymbol{\Omega}$ and $\boldsymbol{\Pi}$ with respect to its symmetry etc., there is no guarantee that the residual associated with the faulty sensor is the largest one. The above equation also suggests the possibility that variable contributions which, by default, should be insignificant may, in fact, be significant and indicate that these variables are affected by the fault, which this simulation example confirms. For $d = m = m^*$, Eqs. (4.19) and (4.20) suggest that $\Delta I^2{}_{cont,j} = \Delta z_j$, which produces a correct variable contribution to the fault condition.

With the application of the reconstruction technique, a different picture emerges. A total of six fault directions, one for each sensor, were considered. Figure 4.1a shows the fault reconstruction index for each of the examined fault conditions. As expected, a bias for the second sensor yielded the smallest value and below 1 and hence, according to Eq. (4.14) the most significant impact of the reconstruction approach is for this sensor. In contrast, reconstructing the other fault directions (sensor faults) had an insignificant impact, leaving the fault reconstruction indices to exceed 1. It should be noted that plots a–c represent the average values for the 250 faulty samples. For illustrative purposes, Fig. 4.1 also shows the reconstruction indices and the variable contributions for the first ten samples in plots d–f.

Next, we consider a multidimensional fault case. A multiple sensor fault is generated by injecting sensor bias of magnitude 2 to sensors 2 and 5 after the 250th sample of the test dataset. Along with the six fault conditions describing a bias for each sensor, there are now seven conditions and consequently seven different fault directions where the direction for the multidimensional sensor bias is

$$\boldsymbol{\Theta}_7 = \begin{bmatrix} 0 & 1 & 0 & 0 & 0 & 0 \\ 0 & 0 & 0 & 0 & 1 & 0 \end{bmatrix}^T \tag{4.21}$$

Figure 4.2 summarizes the results of analyzing the 250 faulty samples. Plot a in this figure confirms that this event was correctly diagnosed as fault condition no. 7. In contrast, none of the remaining fault conditions produced a significant effect upon the corresponding fault identification index and hence produced values that exceeded 1. As before, plots b and c show that contribution charts for the ICA model, which again, produce a misleading picture. This can be explained through the use of Eqs. (4.17) to (4.20). Plots d–f show the reconstruction indices and the variable contribution charts for the first ten samples, respectively.

4.3 Fault Identification

4.3.1 PCA-Based Similarity Factor

Krzanowski (1979) developed a method for measuring the similarity of two datasets using a PCA similarity factor, S_{PCA}. Let us consider two datasets, $\mathbf{D}_1$ and $\mathbf{D}_2$ that have the same variables but not necessarily the same number of measurements. Suppose we have selected k principal components for the two PCA models. The PCA similarity factor is a measure of the similarity between datasets $\mathbf{D}_1$ and $\mathbf{D}_2$ and to be exact, it is between the two reduced subspaces. Denoted $\mathbf{L}$ and $\mathbf{M}$, the two subspaces, the PCA similarity factor can be calculated from the angles between principal components

$$S_{PCA} = \frac{1}{k}\sum_{i=1}^{k}\sum_{j=1}^{k}\cos^2\Theta_{ij} \tag{4.22}$$

where Θ_{ij} is the angle between the ith principal component of dataset $\mathbf{D}_1$ and the jth principal component of dataset $\mathbf{D}_2$ which is defined as $\cos\Theta_{ij} = \frac{\mathbf{L}_i^T\mathbf{M}_j}{||\mathbf{L}_i||_2||\mathbf{M}_j||_2}$, here $||\mathbf{L}_i||_2 = ||\mathbf{M}_j||_2 = 1$ and $0 \leq \Theta \leq \pi/2$ S_{PCA} can also be expressed by subspaces $\mathbf{L}$ and $\mathbf{M}$ as

$$S_{PCA} = \frac{trace(\mathbf{L}^{\mathrm{T}}\mathbf{M}\mathbf{M}^{\mathrm{T}}\mathbf{L})}{k} \tag{4.23}$$

If the value of S_{PCA} exceeds a specified cutoff, the two datasets $\mathbf{D}_1$ and $\mathbf{D}_2$ are considered to be similar. By introducing different weight for each principal component, this similarity factor has been modified by Singhai and Seborg (2002). Besides, Zhao et al. (2004) also modified the similarity factor by introducing the principal angle, in order to mine the most similar information between two datasets.

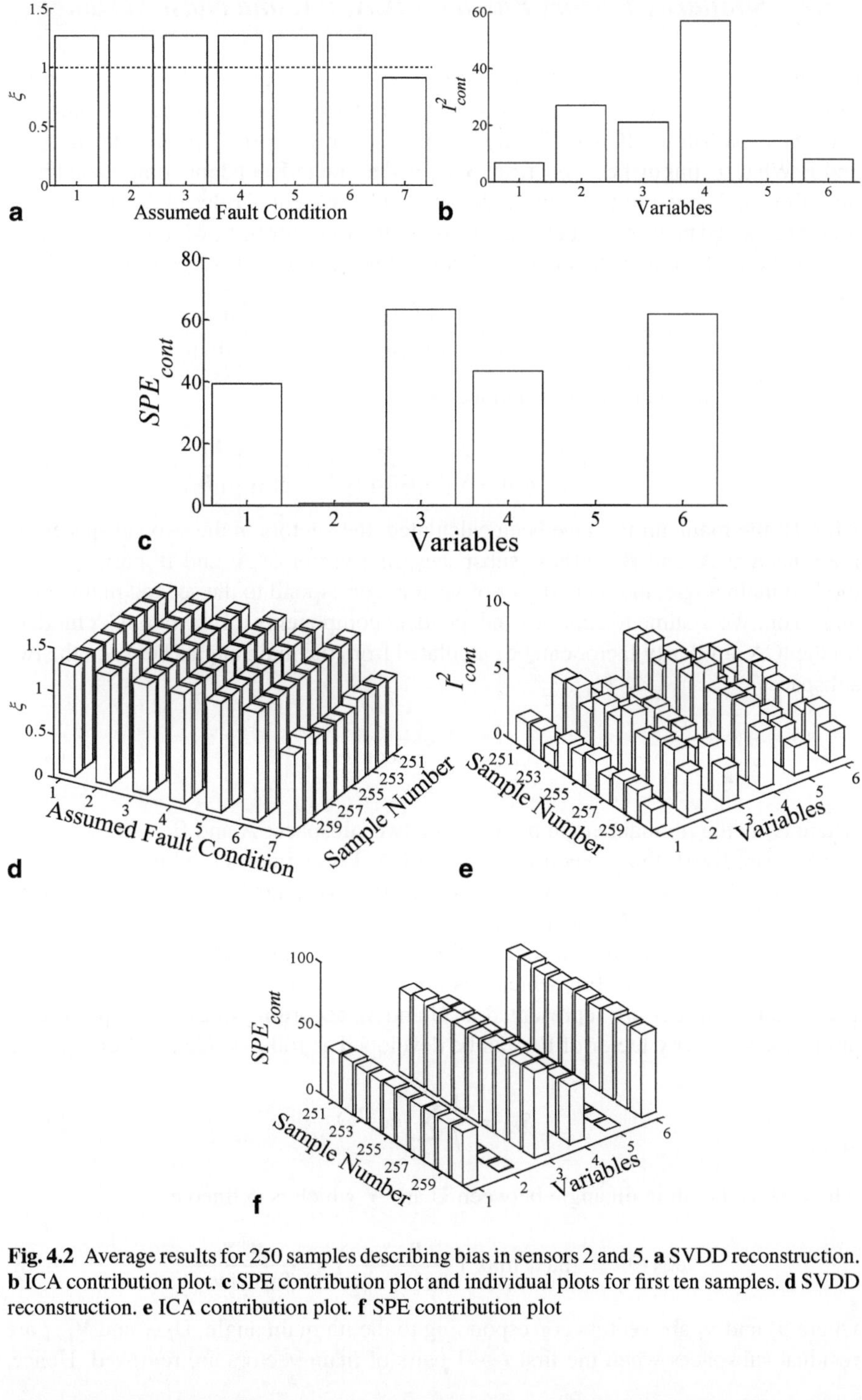

Fig. 4.2 Average results for 250 samples describing bias in sensors 2 and 5. **a** SVDD reconstruction. **b** ICA contribution plot. **c** SPE contribution plot and individual plots for first ten samples. **d** SVDD reconstruction. **e** ICA contribution plot. **f** SPE contribution plot

4.3.2 Similarity Factors Based on ICA, FA, and Noise Variance

However, the PCA similarity factor is just a measure of the similarity between the Gaussian part of datasets $\mathbf{D}_1$ and $\mathbf{D}_2$ For the similarity measure of the non-Gaussian part, we establish the ICA similarity factor S_{ICA} for the corresponding subspaces A and B What distinguished S_{ICA} from S_{PCA} is that the independent components have no orthogonal constraint, and these extracted ICs have equal roles in the ICA model. So we proposed new angles between the two subspaces for ICA, which we call "main angles" here, these new "main angles" are defined as follows (Ge and Song 2007)

$$\cos\Theta_1 = \max_{\mathbf{a}\in\mathbf{A}}\max_{\mathbf{b}\in\mathbf{B}}\left(\frac{\mathbf{a}^T\mathbf{b}}{||\mathbf{a}||_2||\mathbf{b}||_2}\right) = \frac{\mathbf{a}_1^T\mathbf{b}_1}{||\mathbf{a}_1||_2||\mathbf{b}_1||_2} \tag{4.24}$$

Hence, the ith main angle Θ_i is defined as

$$\cos\Theta_i = \max_{\mathbf{a}\in\mathbf{A}_{i-1}}\max_{\mathbf{b}\in\mathbf{B}_{i-1}}\left(\frac{\mathbf{a}^T\mathbf{b}}{||\mathbf{a}||_2||\mathbf{b}||_2}\right) = \frac{\mathbf{a}_i^T\mathbf{b}_i}{||\mathbf{a}_2||_i||\mathbf{b}_i||_2} \tag{4.25}$$

After all the main angles have been calculated, the vectors in the two subspaces are rearranged as $\mathbf{A}'$ and $\mathbf{B}'$ In these subspaces, first vector of $\mathbf{A}'$ and $\mathbf{B}'$ correspond to the first main angle, the second pair of vectors correspond to the second main angle, and so on. We assume to choose r independent components for the two ICA models. So the ICA similarity factor can be calculated from the main angles between the two subspaces

$$S_{ICA} = \frac{1}{r}\sum_{i=1}^{r}\cos^2\Theta_i \tag{4.26}$$

where Θ_i is the ith main angle between the two subspaces A and B.

Compared with the projection method PCA, FA estimates the loading matrix $\mathbf{P}$ in the probability space. The loading matrix $\mathbf{P}_{pca}$ is ranged due to the value of its corresponding eigenvalue, and the matrix is restricted that $\mathbf{P}_{pca}^T\mathbf{P}_{pca} = \mathbf{I}$. However, the new estimated loading matrix $\mathbf{P}$ of FA has no such properties. Besides, the noise part is also estimated in FA, which has not been considered in PCA. Suppose two FA loading matrices are represented as $\mathbf{U}$ and $\mathbf{V}$, according to the concept of main angle, the similarity factor of FA can be developed as follows (Ge and Song 2009a)

$$S_{FA} = \frac{1}{l}\sum_{i=1}^{l}\cos^2\Theta_i \tag{4.27}$$

where Θ_i is the ith main angle between $\mathbf{U}$ and $\mathbf{V}$ which is defined as

$$\cos\Theta_i = \max_{\mathbf{u}\in\mathbf{U}_{i-1}}\max_{\mathbf{v}\in\mathbf{V}_{i-1}}\left(\frac{\mathbf{u}^T\mathbf{v}}{\|\mathbf{u}\|_2\|\mathbf{v}\|_2}\right) = \frac{\mathbf{u}_i^T\mathbf{v}_i}{\|\mathbf{u}_i\|_2\|\mathbf{v}_i\|_2} \tag{4.28}$$

where $\mathbf{u}_i$ and $\mathbf{v}_i$ are vectors corresponding to the ith main angle, $\mathbf{U}_{i-1}$ and $\mathbf{V}_{i-1}$ are residual subspaces when the first $i-1$ pairs of main vectors are removed. Hence, the most similar information between the two datasets can be extracted by FA-based similarity factor S_{FA}.

Next, we consider the similarity between the noise parts of the two datasets. Suppose the estimated noise matrices of the two fault datasets are $\boldsymbol{\Sigma}_1$ and $\boldsymbol{\Sigma}_2$ Due to the diagonal property of the noise covariance matrix, the diagonal elements can be extracted, thus

$$\begin{aligned}\boldsymbol{\lambda}_1 &= diag(\boldsymbol{\Sigma}_1)\\ \boldsymbol{\lambda}_2 &= diag(\boldsymbol{\Sigma}_2)\end{aligned} \tag{4.29}$$

where $\boldsymbol{\lambda}_1$ and $\boldsymbol{\lambda}_2$ are two vectors, which consist of diagonal elements of $\boldsymbol{\Sigma}_1$ and $\boldsymbol{\Sigma}_2$ The similarity factor between the noise parts of two fault patterns can be defined as (Ge and Song 2009a)

$$\begin{aligned}S_N &= \sqrt{e^{-d^2(\boldsymbol{\lambda}_1,\boldsymbol{\lambda}_2)}}\\ d(\boldsymbol{\lambda}_1,\boldsymbol{\lambda}_2) &= ||\boldsymbol{\lambda}_1,\boldsymbol{\lambda}_2||_2\end{aligned} \tag{4.30}$$

In summary, there are three similarity factors S_{ICA}, S_{FA} and S_N defined for similarity measurement of two fault datasets. When these three similarity factors are used simultaneously, a weighted average of these similarity factors can be defined as follows

$$S_{avg} = \alpha \cdot S_{ICA} + \beta \cdot S_{FA} + \gamma \cdot S_N \tag{4.31}$$

where $0 \leq \alpha, \beta, \gamma \leq 1$ and $\alpha + \beta + \gamma = 1$, they are weighted factors. How to choose theses weighted factors is an open question, which means they should be determined accordingly under different situations. However, in our experience, α and β can be chosen bigger than γ, because the systematic part takes more responsibility that two datasets differentiated from each other. The choices of the three weighted factors can be determined as

$$\begin{aligned}\frac{\alpha}{\beta} &= \frac{r}{l}\\ \gamma &= \frac{1}{2}\mathbf{min}\{\alpha, \beta\}\end{aligned} \tag{4.32}$$

where r and l are numbers of non-Gaussian and Gaussian essential variables. Therefore, if $r \geq l$, Eq. (4.31) becomes

$$S_{avg} = \frac{2r}{2r+3l} \cdot S_{ICA} + \frac{2l}{2r+3l} \cdot S_{FA} + \frac{l}{2r+3l} \cdot S_N. \tag{4.33}$$

Else, if $r < l$, Eq. (4.31) becomes

$$S_{avg} = \frac{2r}{3r+2l} \cdot S_{ICA} + \frac{2l}{3r+2l} \cdot S_{FA} + \frac{r}{3r+2l} \cdot S_N. \tag{4.34}$$

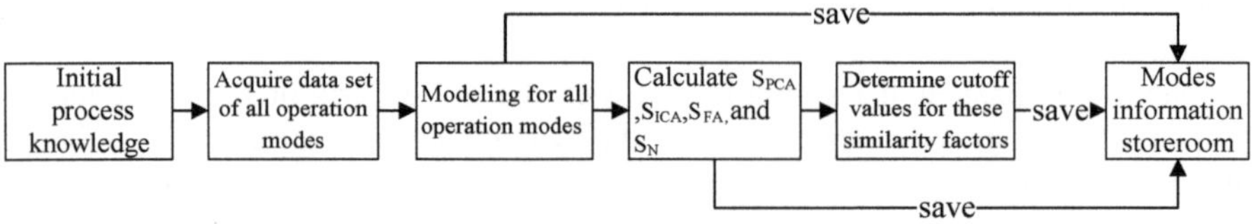

Fig. 4.3 Modes information storeroom development

4.3.3 Implementation Procedures of Fault Identification Based on Similarity Factors

Before identifying a detected fault, a modes information storeroom should be built. This storeroom contains all the operation modes (normal and fault) information of the process. The procedures of building the modes information storeroom are given as follows

Step 1 Acquire data set **X** for all operation modes of the process.
Step 2 For each operation mode, calculate its mixing matrix **A**, load matrix **P**, and the noise variance.
Step 3 Calculate similarity factor values for every pair of the operation modes, including $\mathbf{S}_{PCA}$, $\mathbf{S}_{ICA}$, $\mathbf{S}_{FA}$ and $\mathbf{S}_{N}$.
Step 4 Determine cutoff values for these similarity factors.
Step 5 Preserve all the parameters and matrices in the modes information storeroom.

The systematic flowchart of building the modes information storeroom is shown as follows (Fig. 4.3).

When a mode storeroom has been built, the mode identification for the detected fault can be carried out through the following procedures

Step 1 Acquire dataset $\mathbf{X}_{fault}$ for the fault operation mode of the process.
Step 2 Calculate its mixing matrix $\mathbf{A}_{fault}$, load matrix $\mathbf{P}_{fault}$, and λ_{fault}.
Step 3 Calculate similarity factor values, which is between the fault operation mode and every mode in the storeroom.
Step 4 If the similarity factor value exceeds the cutoff value, the detected fault mode is considered as the corresponding operation mode in the storeroom.
Step 5 If none of the similarity factor value exceeds the cutoff value, the detected fault mode is considered as a new operation mode, and the storeroom should be updated.

The mode identification scheme of the similarity factor-based method is shown below (Fig. 4.4).

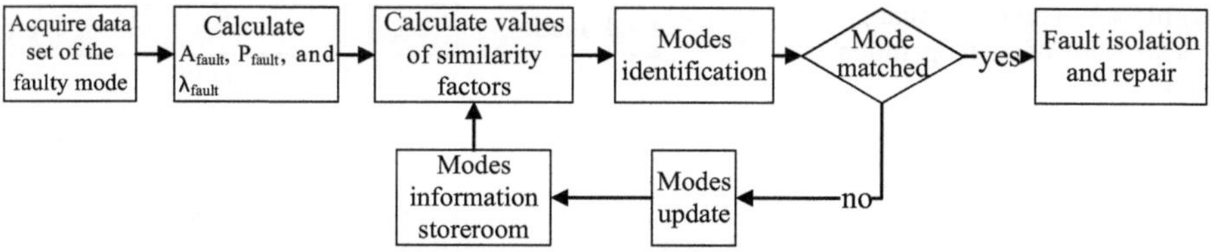

Fig. 4.4 Mode identification strategy for the similarity factor-based method

Table 4.1 Monitoring variables in the TE process

No.	Measured variables
1	A feed
2	D feed
3	E feed
4	A and C feed
5	Recycle flow
6	Reactor feed rate
7	Reactor temperature
8	Purge rate
9	Product separator temperature
10	Product separator pressure
11	Product separator underflow
12	Stripper pressure
13	Stripper temperature
14	Stripper steam flow
15	Reactor cooling water outlet temperature
16	Separator cooling water outlet temperature

4.3.4 Case Study of TE Benchmark Process

In this section, the performance of the fault identification method is tested through the TE process. The detailed description of the TE process can be found in Chap. 3, and also in the book by Chiang et al. (2001). The variables selected for monitoring are the same as that Kano et al. (2002), which are listed in Table 4.1. Simulation data which are acquired in the normal process consist of 960 samples and their sampling interval was 3 min. In order to evaluate fault identification performance of the similarity factor-based methods, simulation data under 21 fault processes are also obtained.

Following the modeling procedures, signature of each dataset is calculated, thus the parameter set $\{\mathbf{A}_{new}, \mathbf{P}_{new}, \mathbf{\Sigma}_{new}\}_{new=0,1,\cdots 21}$ is obtained. Then, the similarity between the new signatures (operation modes) and the preserved signatures are calculated through Eq. (4.34). According to Ge and Song (2007), the similarity values of S_{mix} (the mixed similarity factor) between different operating modes are smaller compared to that of $S_{PCA}$and S_{CS} (the similarity factor use the first principal vector), indicating that operating modes are different from each other. Successful identification depends on the discrimination potential of the disturbance modes as high similarity between modes would limit the potential to distinguish between the

Table 4.2 Average similarity factor value of each operation mode

Fault mode	S_{avg}	S_{mix}	S_{PCA}
0	*0.544*	0.695	0.850
1	*0.583*	0.622	0.824
2	*0.393*	0.672	0.815
3	*0.734*	0.830	0.874
4	*0.633*	0.658	0.835
5	*0.616*	0.670	0.864
6	*0.377*	0.567	0.671
7	*0.539*	0.598	0.854
8	*0.613*	0.691	0.826
9	*0.567*	0.751	0.873
10	0.715	*0.694*	0.879
11	0.709	*0.693*	0.851
12	*0.443*	0.680	0.841
13	*0.591*	0.715	0.821
14	*0.520*	0.700	0.831
15	0.767	*0.705*	0.872
16	*0.700*	0.733	0.876
17	*0.421*	0.668	0.844
18	*0.364*	0.616	0.692
19	0.781	*0.677*	0.833
20	*0.611*	0.674	0.850
21	0.737	*0.638*	0.810
Average	*0.589*	0.679	0.831

corresponding faults. So the low similarity values of S_{mix} show power for mode identification. To compare the values of different similarity factors, the average of each fault identification value is tabulated in Table 4.2, including similarity factors S_{avg}, S_{mix}, and S_{PCA}. The minimum value achieved for each mode is marked with a bold number. The smaller the similarity factor value achieved the more powerful of the discrimination potential the corresponding similarity factor shows. Therefore, the minimum value of the similarity factor means the best fault identification performance of the corresponding similarity factor. As indicated in Table 4.2, most of the minimum values are achieved by the S_{avg} similarity factor, whilst only five operation modes show the superiority of the S_{mix} similarity factor. In comparison, S_{PCA} gives the worst fault identification performance. Besides, the average values of all the 22 operation modes are also tabulated, in which the best average performance is also achieved by the S_{avg} similarity factor.

4.4 Conclusions

In this chapter, a non-Gaussain fault reconstruction scheme has been demonstrated, which is based on the ICA and SVDD model. Different from the PCA model-based Gaussian fault reconstruction method, this scheme is capable of handling the non-Gaussian distribution of the process data. Compared to the traditional contribution-based method, more accurate fault diagnosis results have been obtained by the non-Gaussian reconstruction method. For identification of the fault type, the PCA-based similarity factor has been extended to the ICA similarity factor and the noise similarity factor, based on which a combined similarity factor has also been developed. Through an application case study of the TE benchmark process, the effectiveness and superiority of the similarity factors have been confirmed for fault identification.

4.5 Appendix: Iterative Reconstruction Procedure in the Feature Space

Section 4.2 showed that **z** was reconstructed without the consideration of noises and outliers. To develop a robust iterative reconstruction scheme, Eq. (4.9) can be rewritten as follows

$$\mathbf{f}(t+1) = ({\Theta_j}^T \mathbf{W}^T \mathbf{W} \Theta_j)^{-1} \frac{{\Theta_j}^T \mathbf{W}^T \sum_{i=1}^{K} \alpha_i K(\mathbf{sv}_i, \boldsymbol{\tau}(t))(\hat{\mathbf{s}} - \mathbf{sv}_i)}{\sum_{i=1}^{K} \alpha_i K(\mathbf{sv}_i, \boldsymbol{\tau}(t))} \tag{4.35}$$

where

$$\boldsymbol{\tau}(t) = \begin{cases} \hat{\mathbf{s}} & t = 0 \\ \hat{\mathbf{s}} - \mathbf{W}\Theta_j \mathbf{f}(t-1) & t > 0 \end{cases} \tag{4.36}$$

However, according to Takahashi and Kurita (2002), above updating procedure may fail during the iteration unless $\tau(0)$ locates in the near neighborhood of $\hat{\mathbf{s}}^*$. To prevent this issue, we introduce a robust form of the update in Eq. (4.29), which is given by

$$\boldsymbol{\tau}(t) = \begin{cases} \hat{\mathbf{s}} & t = 0 \\ \mathbf{B}(t)\hat{\mathbf{s}} + [\mathbf{I} - \mathbf{B}(t)](\hat{\mathbf{s}} - \mathbf{W}\Theta_j \mathbf{f}(t-1)) & t > 0 \end{cases} \tag{4.37}$$

where **I** is a $d \times d$ identity matrix and the $d \times d$ matrix $\mathbf{B}(t)$ is defined as diag$\{b_1(t), \ldots, b_d(t)\}$. Each element $\mathbf{B}(t)$ denotes the "certainty" of $\hat{\mathbf{s}}$ and can be estimated by the difference between **sv** and its corresponding reconstruction $\hat{\mathbf{s}}(t-1) = \mathbf{s} - \mathbf{W}\Theta_j \mathbf{f}(t-1)$

$$b_i(t) = \exp\left(-\frac{1}{2}\frac{-[\hat{\mathbf{s}}_i - \hat{\mathbf{s}}_i(t-1)]^2}{\sigma_i^2}\right) \tag{4.38}$$

where $i = 1, 2, \ldots d$ and the parameter σ_i is the standard deviation of the differences, given by (Takahashi and Kurita 2002)

$$\sigma_i = 1.4826\left(1 + \frac{5}{K-1}\right) \text{med}_j \sqrt{\varepsilon_{ji}^2} \tag{4.39}$$

where ${\varepsilon^2}_{ji}$ is the squared error between the ith component of the jth support vector and the reconstructed value, and $\mathrm{med}_j\sqrt{\varepsilon_{ji}^2}$ is the median of $\sqrt{\varepsilon_{ji}^2}$. The next iteration implies that the original value **s** is used for the next step if it has a high "certainty." Otherwise, the reconstructed value $\hat{\mathbf{s}}(t-1)$ is used as its estimation. The iteration has converged if $\|\boldsymbol{\tau}(t)-\boldsymbol{\tau}(t-1)\| < e$ or if $\|\mathbf{f}(\mathrm{t})-\mathbf{f}(t-1)\| < e$ with e being the convergence criterion, e.g., 10^{-5}.

Chapter 5
Nonlinear Process Monitoring: Part 1

5.1 Introduction

Multivariate statistical process control (MSPC) has been widely used in monitoring of large-scale industrial processes (Kresta et al. 1991; Nomikos and MacGregor 1994, 1995a, b; Bakshi 1998; Dunia and Qin 1998; Kano et al. 2002; Singhai and Seborg 2006). Traditionally, MSPC employs principal component analysis (PCA) for describing the relationship between process variables. PCA is a dimensionality reduction technique that can handle high-dimensional and correlated data. It separates data information into two parts: a systematic part and a noisy part. The systematic part contains most variation in the data, while the noisy part has least variances. For process monitoring, PCA used two statistics, represented by Mahalanobis and Euclidean distances to detect changes in the systematic part and the noisy one, respectively. However, for some complicated cases in industrial chemical and biological processes with particular nonlinear characteristics, PCA performs poorly due to its assumption that the process data are linear (Lee et al. 2004).

To overcome the nonlinear shortcoming of the traditional MSPC method, several nonlinear extensions of PCA were reported. Kramer (1992) developed a nonlinear PCA method based on the autoassociative neural network. Dong and McAvoy (1996) proposed a nonlinear PCA by combining the principal curve and the neural network. After obtaining the associated scores and the correlated data using the principal curve method, they used a neural network model to map the original data into scores and to map these scores into the original variables. Alternative nonlinear PCA methods including Hiden et al. (1999), Cheng and Chiu (2005), Zhao and Xu (2005), and Maulud et al. (2006) have also been proposed to solve the nonlinear process monitoring problem. Kruger et al. (2005) proposed a new nonlinearity measure for principal component and discussed whether the process can be described by linear PCA or whether the nonlinear PCA should be used. However, most of the existing nonlinear PCA approaches are based on neural network, which has to solve a nonlinear optimization problem.

Recently, a new nonlinear process monitoring method based on kernel PCA (KPCA) has been proposed (Lee et al. 2004; Choi et al. 2005; Cho et al. 2005). The main idea of KPCA is first to map the input space into a feature space via a

Z. Ge, Z. Song, *Multivariate Statistical Process Control,* Advances in Industrial Control,
DOI 10.1007/978-1-4471-4513-4_5, © Springer-Verlag London 2013

nonlinear map, which makes data structure more linear, and then to extract principal components in the feature space. By introducing a kernel function, the nonlinear mapping and the inner product computation can be avoided. The main advantage of the KPCA method over referred nonlinear PCA approaches is that no nonlinear optimization should be involved. Similar to the conventional PCA, two monitoring statistics have been constructed to monitor the systematic and the noisy part separately. However, the confidence limit determination of the systematic monitoring statistic T^2 is made under the assumption that the obtained score variable follows Gaussian distribution. This assumption is met for linear processes. But for nonlinear processes, it can no longer be assumed. Due to the nonlinear transformation, extracted score variables can hardly follow Gaussian distribution. Therefore, the score variable can no longer be used to construct the T^2 statistic for monitoring and also the confidence limit based on a predefined parametric probability density function cannot be determined through the known F distribution. Besides, the SPE statistic may also have the same problem as T^2 statistic. If the variable is not Gaussian distributed, the confidence limit of the SPE statistic is not held any more.

Statistical local approach is a novel method for abrupt change detection (Basseville and Nikiforov 1993). This method is based on monitoring changes in model parameters of a given function which describes the system. A parameter departure can be identified by a change in the mean value of a Gaussian probability density function. Zhang et al. (1994) extended the local approach to incorporate model reduction, biased identification, and even the simple Monte-Carlo estimation technique can be involved in. Basseville (1998) used this method to perform the on-board incipient fault detection and isolation tasks which are involved in monitoring of condition-based maintenance. Recently, the statistical local approach has been introduced into PCA to detect incipient changes in the variable covariance structure (Kruger et al. 2007). However, the method was only restricted in the linear case and detection of incipient changes. No improvement of data distributed restriction has been made to put into practice.

To address the problems stated above, this chapter introduces the statistical local approach (Basseville 1998) into KPCA for nonlinear process monitoring. With the incorporation of the novel statistical local approach, whatever the initiative distribution of the score variable, the new constructed statistic from the local approach follows a Gaussian distribution. Besides, the confidence limit of the new statistic can be easily determined by a χ^2 distribution (Ge and Song 2009).

5.2 Incorporation of Statistical Local Approach and KPCA

5.2.1 Statistical Local Approach

The statistical local approach is perfectly described by Basseville (1998). Denote θ and θ_0 as parameters that represent abnormal and normal behaviors of the process.

Using the statistical local approach, the abnormal parameter can be written as

$$\theta = \theta_0 + \frac{\Delta\theta}{\sqrt{n}} \tag{5.1}$$

where $\Delta\theta$ is a fixed but unknown vector. The hypothesis test is given as follows

$$H_0 : \theta = \theta_0, \quad H_1 : \theta = \theta_0 + \frac{\Delta\theta}{\sqrt{n}} \tag{5.2}$$

The central limit theorem shows that the probability density function of H_1 hypothesis asymptotically follows a Gaussian distribution

$$\begin{aligned} \zeta &\sim N(\mathbf{M}(\theta_0)\Delta\theta, \Sigma(\theta_0)) \\ \mathbf{M}(\theta_0) &\triangleq -\frac{\partial\psi(\theta, \mathbf{x}(j))}{\partial\theta}\bigg|_{\theta=\theta_0} \end{aligned} \tag{5.3}$$

where ζ refers to the improved residual vector, which is defined in Eq. (5.4) and $\psi(\theta, \mathbf{z}(j))$ is the primary residual (Basseville 1998)

$$\zeta \triangleq \frac{1}{\sqrt{n}} \sum_{j}^{n} \psi(\theta_0, \mathbf{x}(j)) \tag{5.4}$$

According to Basseville (1998), the primary residual should follow assumptions given as below

1. $E\{\psi(\theta_0, \mathbf{x}(j))\} = 0$ if $\theta = \theta_0$.
2. $E\{\psi(\theta, \mathbf{x}(j))\} \neq 0$ if $\theta \in \omega(\theta_0), \theta_0 \notin \omega(\theta_0)$.
3. $\psi(\theta_0, \mathbf{x}(j))$ is differentiable in θ.
4. $\psi(\theta, \mathbf{x}(j))$ exists in the vicinity of θ_0, thus $\theta \in \omega(\theta_0)$.

5.2.2 Introducing Statistical Local Approach into KPCA Monitoring Framework

The statistical local approach is based on formulating changes of parameters which represent the process behavior. The new method focuses on the change of kernel eigenvectors or eigenvalues. The method was first used by Kruger (2007) in linear PCA method for incipient change monitoring. According to Basseville (1998), it is important to choose a monitoring function that could derive sufficient primary residuals. For linear PCA, monitoring functions for eigenvector and eigenvalue changes both have been proposed by Kruger (2007). Sufficiency of the primary residual constructed from these monitoring functions has also been proved (Kruger et al. 2007). This section derives required monitoring functions in KPCA for the change of kernel eigenvectors and eigenvalues. Primary residuals are developed, followed by the proof of their sufficiency for process monitoring.

Similar to PCA, the objective of KPCA to extract a principal component relies on the following optimization function

$$\begin{aligned}\mathbf{v}_i &\triangleq \arg\max_{\mathbf{v}}\{\mathbf{E}(\mathbf{v}^T\bar{\Phi}(\mathbf{x})\bar{\Phi}(\mathbf{x})^T\mathbf{v}) - \lambda_i(\mathbf{v}^T\mathbf{v}-1)\} \\ &= \arg\max_{\alpha}\{\mathbf{E}(\alpha^T\bar{\mathbf{K}}(\mathbf{x})\bar{\mathbf{K}}(\mathbf{x})^T\alpha) - \lambda_i(\alpha^T\bar{\mathbf{K}}\alpha-1)\}\end{aligned} \tag{5.5}$$

where $\mathbf{E}(\cdot)$ is the statistical expectation operator, $\mathbf{v}$ is the kernel eigenvector, and α is a coefficient vector. The $\mathbf{E}(\mathbf{v}^T\bar{\Phi}(\mathbf{x})\bar{\Phi}(\mathbf{x})^T\mathbf{v})$ represents the variance of the ith kernel score variable, and λ_i is the corresponding Lagrangian multiplier (kernel eigenvalue). The solution to Eq. (5.5) is given by (Ge and Song 2009)

$$\alpha_i \triangleq \arg\frac{\partial}{\partial\alpha_i}\{\mathbf{E}(\alpha^T\bar{\mathbf{K}}(\mathbf{x})\bar{\mathbf{K}}(\mathbf{x})^T\alpha) - \lambda_i(\alpha^T\bar{\mathbf{K}}(\mathbf{x})\alpha-1)\} = 0 \tag{5.6}$$

which becomes as

$$\alpha_i \triangleq \arg\{\mathbf{E}(2\mathbf{t}_i\bar{\mathbf{K}}(\mathbf{x})) - 2\lambda_i\bar{\mathbf{K}}(\mathbf{x})\alpha_i\} = 0 \tag{5.7}$$

Define

$$\Omega_i = 2\mathbf{t}_i\bar{\mathbf{K}}(\mathbf{x}) - 2\lambda_i\bar{\mathbf{K}}(\mathbf{x})\alpha_i \tag{5.8}$$

Equation (5.7) can be represented as

$$\alpha_i \triangleq \arg\{\mathbf{E}(\Omega_i)\} = 0 \tag{5.9}$$

Thus

$$\mathbf{E}(\Omega_i)|_{\alpha=\alpha_i} = 0 \tag{5.10}$$

which satisfies assumption (1) given in Sect. 5.2.1 for the primary residual. Other three assumptions can be easily satisfied according to Basseville (1998) and Kruger (2007). Therefore, $\mathbf{\Omega}$ can be used as the monitoring function for monitoring the change in kernel eigenvectors and the primary residual can be calculated from this monitoring function.

Similarly, the monitoring function for the eigenvalue change monitoring can also be developed. Premultiplying Eq. (5.8) by α_i^T, the equation can be written as

$$\alpha_i^T\Omega_i = 2\mathbf{t}_i\alpha_i^T\bar{\mathbf{K}}(\mathbf{x}) - 2\lambda_i\alpha_i^T\bar{\mathbf{K}}(\mathbf{x})\alpha_i \tag{5.11}$$

Denote $\Lambda_i = \alpha_i^T\Omega_i$, and notice $\alpha_i^T\bar{\mathbf{K}}(\mathbf{x})\alpha_i = 1, \mathbf{t}_i = \alpha_i^T\bar{\mathbf{K}}(\mathbf{x})$, Eq. (5.11) becomes as

$$\Lambda_i = 2\mathbf{t}_i^2 - 2\lambda_i \tag{5.12}$$

Four assumptions for the primary residual are also satisfied for Λ. Hence, it can be used as a monitoring function for the kernel eigenvalue change detection, and the corresponding primary residual can be derived from this monitoring function. As highlighted by Kruger (2007), the dimension and computation burden are smaller for Λ compared to $\boldsymbol{\Omega}$ thus Λ is much preferable to be employed for process monitoring.

For monitoring systematic and noisy parts separately, the primary residual of Λ can be partitioned into two parts: Λ_s for the systematic information monitoring and Λ_n for monitoring the noisy information

$$\begin{aligned}\Lambda_s &= \Lambda(1 : n_{kpca}) \\ \Lambda_n&(n_{kpca} + 1 : \dim \mathbf{F})\end{aligned} \tag{5.13}$$

where n_{kpca} is the number of chosen kernel principal component, dim $\mathbf{F}$ is the dimensionality of the feature space. Unlike the linear PCA method, the number of the kernel principal component can be bigger than the process variable number. However, the dimension of feature space is arbitrarily high, thus the confidence limit of the noisy information monitoring statistic may be unrealistically large. To avoid this problem, the dimension of the feature space dim $\mathbf{F}$ can be empirically determined that the cumulative sum of the eigenvalues is above 99 %. For simplicity, the number of dim $\mathbf{F}$ can be determined by omitting eigenvalues smaller than a positive value such as 0.0001.

The sensitivity of monitoring functions or their corresponding primary residuals are investigated with the sufficiency given in Theorem 5.1. The proof of Theorem 5.1 is given in Appendix.

Theorem 5.1 *For detecting changes in systematic and noise parts of nonlinear processes, primary residuals of* Λ_s *and* Λ_n *are sufficient in detecting changes in kernel eigenvalues.*

5.3 Nonlinear Process Monitoring Based on Improved KPCA Method

5.3.1 Process Monitoring Approach

For nonlinear process monitoring of KPCA-based method, two statistics T^2 and *SPE* are developed upon kernel score variables and modeling error, which is similar to the linear PCA method. However, as mentioned in the Introduction, the Gaussian distribution of score variables cannot be ensured for nonlinear processes. Hence, if the T^2 statistic is still used for monitoring changes of the systematic part, the performance may be deteriorated. One can find more detailed description in Sect. 5.4.1. It shows that the Gaussian distributed assumption of the score variable is violated and the T^2 statistic cannot detect any change of the systematic information.

For monitoring of noisy part of the process information, the *SPE* statistic can be used. It is also a measure of goodness of fitness of an observation to the KPCA model. However, the original KPCA method did not provide any method for data reconstruction. Lee et al. (2004) and Cho et al. (2005) have proposed methods for construction of the *SPE* statistic, which is shown as follows

$$SPE = \sum_{j=1}^{\dim \mathbf{F}} t_j^2 - \sum_{j=1}^{n_{kpca}} t_j^2 \tag{5.14}$$

or

$$SPE = \mathbf{k}(\mathbf{x}, \mathbf{x}) - \mathbf{t}^T \mathbf{t} \tag{5.15}$$

However, Similar to the systematic space, with the violation of the Gaussian distributed assumption, the *SPE* statistic is also not efficient to detect changes in the noisy part. To handle this problem, the statistical local approach is incorporated into KPCA, a new statistic is developed as follows (Ge and Song 2009)

$$\zeta(\lambda, \mathbf{x}(j)) \triangleq \frac{1}{\sqrt{n}} \sum_{j}^{n} \psi(\lambda_0, \mathbf{x}(j)) = \frac{1}{\sqrt{n}} \sum_{j}^{n} \Lambda(\lambda_0, \mathbf{x}(j)) \tag{5.16}$$

which is called improved residual by Basseville (1998) and follows Gaussian distribution. It was also highlighted that a small departure of the parameter can be identified by a change in mean value of the Gaussian probability density function. If the process behaves normally, the mean value is 0, otherwise it changes to $\mathbf{M}(\lambda_0)\Delta\lambda$ which is demonstrated in Sect. 5.2.2. However, both Gaussian probability density functions have an equal covariance matrix $\Sigma(\lambda_0)$ (Basseville 1998). Similar to the primary residual Λ, the improved residual ζ can also be divided into two parts, which are given as follows

$$\begin{aligned} \zeta_s &= \zeta(1 : n_{kpca}) \\ \zeta_n &= \zeta(n_{kpca} + 1 : \dim \mathbf{F}) \end{aligned} \tag{5.17}$$

Therefore, the T^2 and *SPE* statistics for monitoring systematic changes and noisy part can be reconstructed as

$$T^2(j) = \zeta_s^T(\lambda, \mathbf{x}(j)) \Sigma_s^{-1}(\lambda_0) \zeta_s(\lambda, \mathbf{x}(j)) \tag{5.18}$$

$$SPE(j) = \zeta_n^T(\lambda, \mathbf{x}(j)) \Sigma_n^{-1}(\lambda_0) \zeta_n(\lambda, \mathbf{x}(j)) \tag{5.19}$$

Confidence limits of two new statistics T^2 and *SPE* can be easily determined by a χ^2 distributed function for which the number of degrees of freedom is equal to the dimension of $\zeta_s(\lambda, \mathbf{x}(j))$ and $\zeta_n(\lambda, \mathbf{x}(j))$. This method was shown to be more sensitive for fault detection by Kruger (2007).

For online process monitoring, a moving window approach can be used to increase the sensitivity of these two statistics and also to reduce computation burden (Zhang et al. 1994; Kruger et al. 2007). The new improved residual is given as follows

$$\zeta(\lambda, \mathbf{x}(j)) \triangleq \frac{1}{\sqrt{w}} \sum_{j=nc-w+1}^{nc} \Lambda(\lambda_0, \mathbf{x}(j)) \tag{5.20}$$

where nc is the current observation number and w is length of the moving window. However, how to choose an appropriate length of moving window is still an open question, which means that the number can only be selected by experience. A large number of w may reduce the number of false alarms, while at the same time it may also reduce the sensitivity in detecting process changes. On the other hand, a small moving window length is more sensitive for change detection but may produce an increase level of false alarms.

5.3.2 *Outline and Implementation*

5.3.2.1 Normal Operating Condition (NOC) Model Development

1. Collect normal operating data $\mathbf{x}_i \in R^m, i = 1, 2, \cdots, n$ and normalize these.
2. Compute the kernel matrix $\mathbf{K} \in R^{n\times n}$, where $\mathbf{K}_{ij} = \Phi(\mathbf{x}_i) \cdot \Phi(\mathbf{x}_j)$, then carry out centering in the feature space using

$$\bar{\mathbf{K}} = \mathbf{K} - \mathbf{1}_n\mathbf{K} - \mathbf{K}\mathbf{1}_n + \mathbf{1}_n\mathbf{K}\mathbf{1}_n \tag{5.21}$$

where $$1_n = \frac{1}{n}\begin{bmatrix} 1 & \cdots & 1 \\ \vdots & \ddots & \vdots \\ 1 & \cdots & 1 \end{bmatrix} \in R^{n\times n}$$

3. Solve the eigenvalue problem $\lambda\alpha = \frac{1}{n}\bar{\mathbf{K}}\alpha$ and normalize α such as $\|\alpha_i\| = 1/(n\lambda_i)$.
4. Determine the number of nonlinear principal components via variance explanation method as used in PCA, thus the number of nonlinear principal component n_{kpca} can be selected as

$$CPV = \frac{sum\{\lambda(1:n_{kpca})\}}{sum\{\lambda\}} > \eta \tag{5.22}$$

where η is a threshold value, which is always chosen as $85\% \le \eta \le 95\%$.
5. For normal operating data $\mathbf{x}$, extract a nonlinear principal component via

$$t_k = \langle \mathbf{v}_k, \bar{\Phi}(\mathbf{x})\rangle = \sum_{i=1}^{n} \alpha_i^k \bar{\mathbf{k}}(\mathbf{x}_i, \mathbf{x}) \tag{5.23}$$

6. Calculate the primary residual via Eq. (5.12) and then the improved residual can be calculated by Eq. (5.20).
7. Calculate the value of monitoring statistics T^2 and SPE of the normal operating data by Eqs. (5.18) and (5.19).
8. Determine confidence limits of T^2 and SPE statistics.

5.3.2.2 Online Monitoring

1. Obtain the new observation data and normalize these.
2. Compute the kernel vector for the new observation by $\mathbf{k}_{new} = [\mathbf{k}_{new,j}(\mathbf{x}_{new}, \mathbf{x}_j)]_{j=1,2,\cdots,n}$, where $\mathbf{x}_j$ are the normal operating data.
3. Center the new kernel vector $\mathbf{k}_{new}$ as follows

$$\bar{\mathbf{k}}_{new} = \mathbf{k}_{new} - \mathbf{1}_{new}\mathbf{K} - \mathbf{k}_{new}\mathbf{1}_n + \mathbf{1}_{new}\mathbf{K}\mathbf{1}_n \tag{5.24}$$

where $\mathbf{K}$ and $\mathbf{1}_n$ can be obtained from the modeling step, and $\mathbf{1}_{new} = \frac{1}{n}[1, \cdots, 1] \in R^{1\times n}$.

4. Extract the nonlinear principal component via

$$t_{new,k} = \left\langle \mathbf{v}_k, \bar{\Phi}(\mathbf{x}_{new}) \right\rangle = \sum_{i=1}^{n} \alpha_i^k \bar{\mathbf{k}}(\mathbf{x}_i, \mathbf{x}_{new}) \tag{5.25}$$

5. Calculate the primary residual via Eq. (5.12) and then the improved residual can be calculated by Eq. (5.20).
6. Calculate the value of monitoring statistics T^2_{new} and SPE_{new} of the new operating data by Eqs. (5.18) and (5.19).
7. Monitor if the T^2_{new} or SPE_{new} statistic exceeds its corresponding confidence limit.

5.4 Case Studies

In this section, we demonstrate the process monitoring performance of the new method by two simulation studies. One is a simple example suggested by Dong and McAvoy (1996), which was also used by Cho et al. (2005) and Lee et al. (2004). The superiority of KPCA is demonstrated by them. The other example is the Tennessee Eastman (TE) benchmark process, which has been widely used for testing of various monitoring approaches. In the present study, we focus the simulation studies on the superiority of the new method, especially compared to KPCA. For KPCA modeling, we use a radial basis function (RBF) kernel $k(\mathbf{x}, \mathbf{y}) = \exp(-\|\mathbf{x} - \mathbf{y}\|^2/\sigma)$. The kernel parameter σ is determined empirically, which is suggested as $\sigma = 5m$ by Cho (2005).

5.4.1 Numerical Example

This simple example contains three variables, which are driven by the same factor

$$\begin{aligned} x_1 &= t + e_1 \\ x_2 &= t^2 - 3t + e_2 \\ x_3 &= -t^3 + 3t^2 + e_3 \end{aligned} \tag{5.26}$$

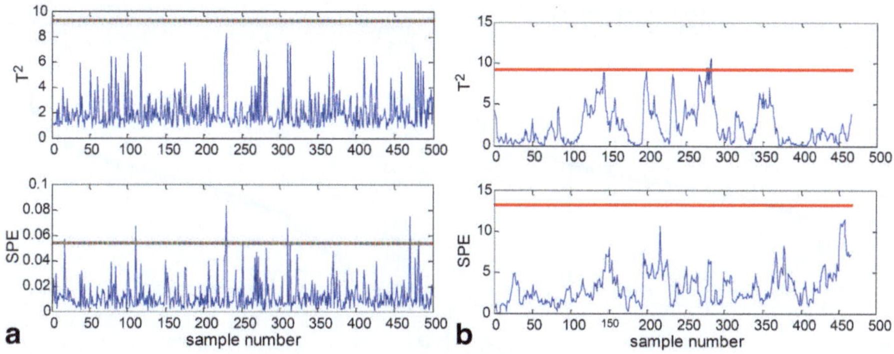

Fig. 5.1 Monitoring results of the process in normal condition. **a** KPCA. **b** New method

where $t \in [0.01, 2]$, and e_1, e_2, and e_3 are independent noise of $N(0, 0.01)$. Five hundred samples were generated by the above equation as the training data. The test data with the sample number of the training data were also generated with the introduction of three artificial faults given as follows

1. Fault 1: A step bias of x_2 by -1 was introduced starting from sample 251.
2. Fault 2: A ramp change of x_1 by adding $0.005(k - 250)$ from the 251th sample.
3. Fault 3: A step change of t by -0.5 was introduced starting from sample 251.

After the training data have been collected, the statistical local approach-based KPCA model and the traditional KPCA model are developed. Different statistics are constructed for process monitoring. A window length $w = 35$ was selected for the establishment of the local approach-based monitoring model. The number of kernel principal component is chosen as 2, which can explain 95.34 % of total variation in the training data. Statistical confidence limits of both methods are set as 99 %. First, monitoring results of the normal process are given in Fig. 5.1, in which good results are given by both methods.

Then, three artificial faults are examined. First two faults have been studied by Cho et al. (2005), in which the T^2 statistic did not detect any process changes. This is because the Gaussian distributed assumption was violated. Precisely, normal probability plots of two extracted kernel principal components are plotted in Fig. 5.2. It is clear that these two principal components are not Gaussian distributed. Therefore, if we still construct the T^2 statistic upon these two non-Gaussian-distributed principal components, the monitoring performance will be deteriorated. Actually, T^2 monitoring results of all three faults are very bad, since it cannot detect any change, which are shown in the fault detection part. In contrast, with the advantage of the statistical local approach, the new statistic (improve residual) will become more Gaussian.

Monitoring results of these three faults are given in Figs. 5.3, 5.4 and 5.5. The first fault is a step change of the second variable which is introduced at sample 251. However, the T^2 statistic of traditional KPCA cannot detect this change, which is also shown by Cho (2005). In comparison, monitoring results of the new method are

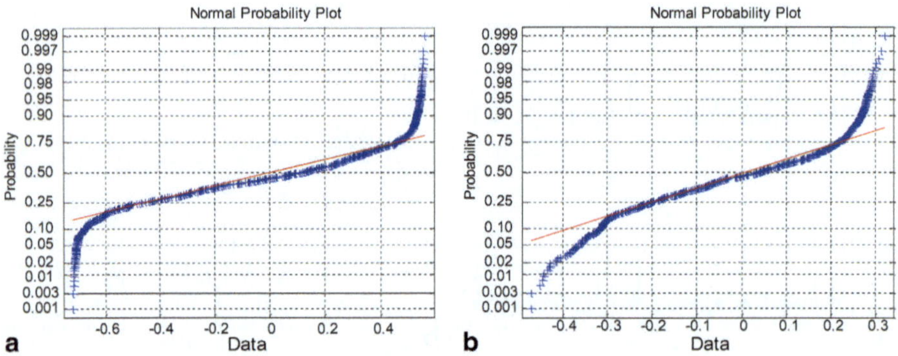

Fig. 5.2 Normal probability plot of kernel principal components. **a** The first kernel principal component. **b** The second kernel principal component

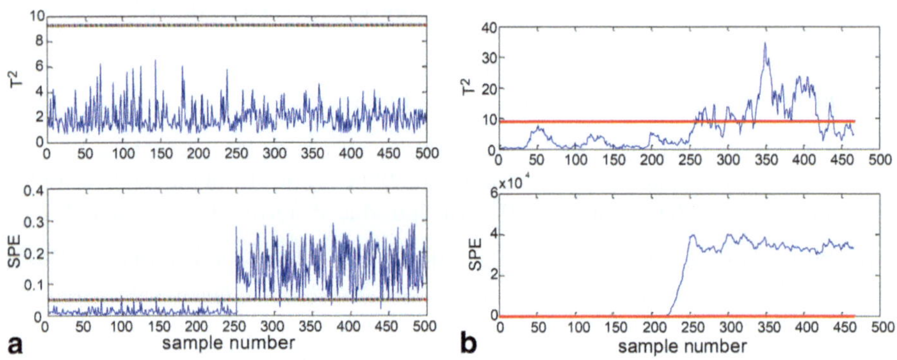

Fig. 5.3 Monitoring results of fault 1. **a** KPCA. **b** New method

much better, since both T^2 and *SPE* statistics can detect this change. Besides,*SPE* is much more sensitive to this change than the corresponding *SPE* statistic of KPCA, since the *SPE* value of the new method in the abnormal condition is extremely large compared to the normal one. If the process changes slowly, such as the second fault, fault detection is difficult especially at the starting period of the change. Monitoring results of this fault are shown in Fig. 5.4. It takes about 50 sample times for KPCA to detect the change. However, the new method can detect the fault immediately when it happens. One may not see it clearly from Fig. 5.4b, since the change of the *SPE* statistic is too rapid. Again, the fault can be detected by the T^2 statistic of the new method, while it is ignored by T^2 of KPCA in Fig. 5.4a. The third fault is a step change of the driven variable *t*, which has not been studied previously by researchers. Monitoring results of KPCA and the new method are shown in Fig. 5.5. The superiority of the new method to KPCA is verified again, both in sensitivity and detection rate.

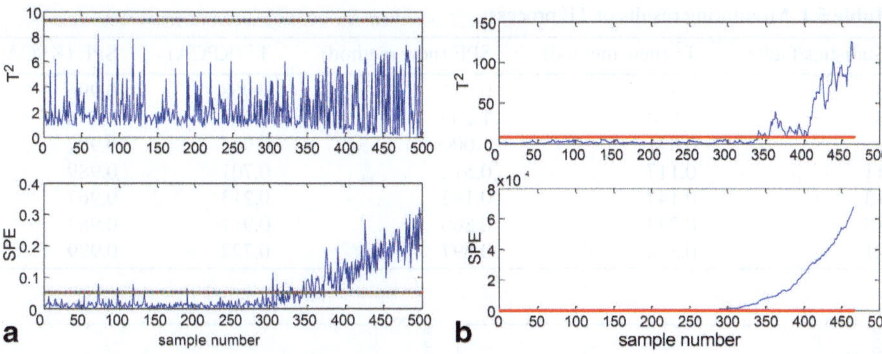

Fig. 5.4 Monitoring results of fault 2. **a** KPCA. **b** New method

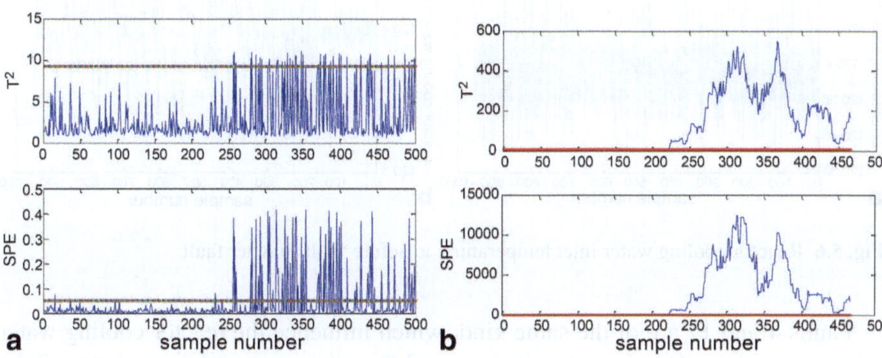

Fig. 5.5 Monitoring results of fault 3. **a** KPCA. **b** New method

5.4.2 TE Benchmark Process

In this section, the TE process is employed for performance evaluation. The detailed description of the TE process can be found in Chap. 3. Variables selected for monitoring are the same as that of Table 4.1.

For monitoring of TE process, 20 kernel principal components are selected, which explain 97.41 % of the total process information. The dimension of feature space is chosen as 26, according to cutoff value of 0.001 for the eigenvalue. Based on the training dataset, two monitoring model of KPCA and the new method are developed. Their corresponding confidence limits (99 %) are also determined. The length of moving window is selected as $w = 70$. Monitoring results (miss detection rate) of some representative faults are tabulated in Table 5.1. The smaller value of the miss detection rate shows the better performance of the corresponding statistic. Detailed demonstration of the superiority of the new method over KPCA is given and analyzed as follows.

Table 5.1 Monitoring results of TE process

Statistics/faults	T^2 (new method)	SPE (new method)	T^2 (KPCA)	SPE (KPCA)
3	0.745	0.994	0.968	0.996
4	0.834	1.000	0.980	0.995
9	0.831	1.000	0.963	0.997
11	0.117	0.512	0.701	0.989
13	0.147	0.182	0.213	0.967
15	0.743	0.860	0.911	0.989
21	0.330	0.697	0.722	0.999

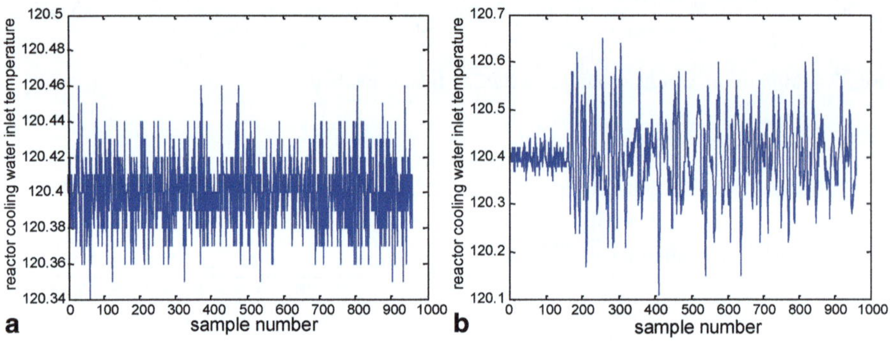

Fig. 5.6 Reactor cooling water inlet temperature. **a** Before fault. **b** After fault

Faults 4 and 11 are of the same kind, which influence the reactor cooling water inlet temperature. One is a step change, and the other is a random variation. Take fault 11 for examination, the reactor cooling water inlet temperature before and after the fault happened are given in Fig. 5.6. From Fig. 5.6b, it is clear that fault 11 is a random variation of this temperature. Monitoring results of both KPCA and the new method are given in Fig. 5.7. Comparing results shown in Fig. 5.7a, 5.7b, the monitoring performance is greatly improved, especially by the T^2 statistic. The random fault is successfully detected after it is introduced into the process, since the T^2 statistic value of the new method keeps above the confidence limit. However, many T^2 statistic values of KPCA stay below its corresponding confidence limit, which means the fault cannot be detected continuously. The monitoring difference between the new method and KPCA may result from the distribution of the score variable. The normality of the score variable is improved after the introduction of the statistical local approach.

To test the sensitivity of the new method, several slow, small, and unknown faults are examined. First, fault 13 is a slow drift fault of the reaction kinetics. Monitoring results of both methods are given in Fig. 5.8. Compared to KPCA, the new method is more sensitive to this fault, since the change of its T^2 statistic value is much more significant than that of KPCA. Figure 5.9 shows monitoring results of fault 15, which is a small fault. This fault causes sticking problem of the condenser cooling water valve. According to Chiang (2000), traditional statistics can hardly detect this

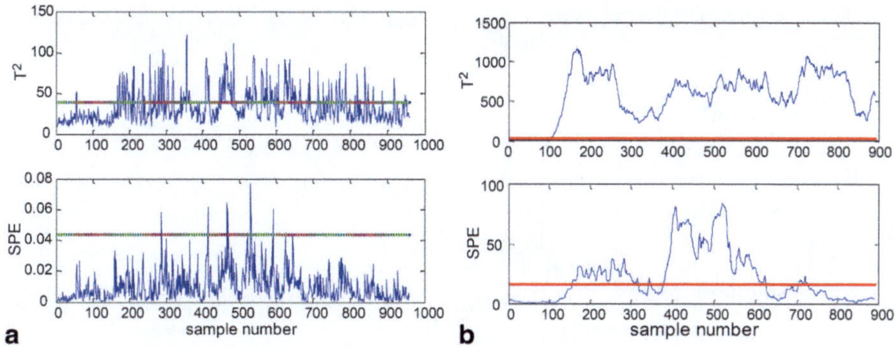

Fig. 5.7 Monitoring results of fault 11. **a** KPCA. **b** New method

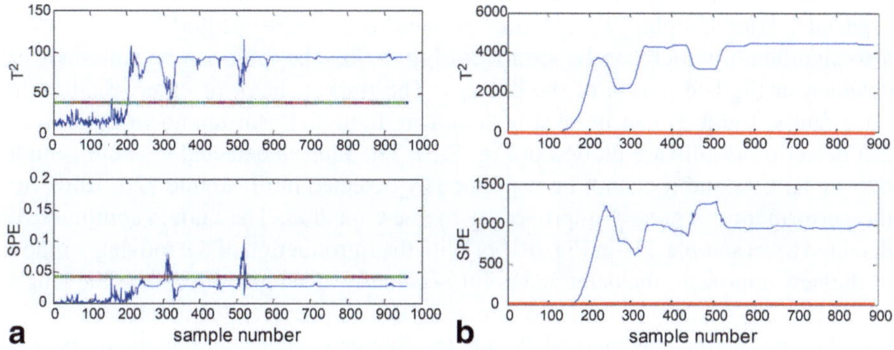

Fig. 5.8 Monitoring results of fault 13. **a** KPCA. **b** New method

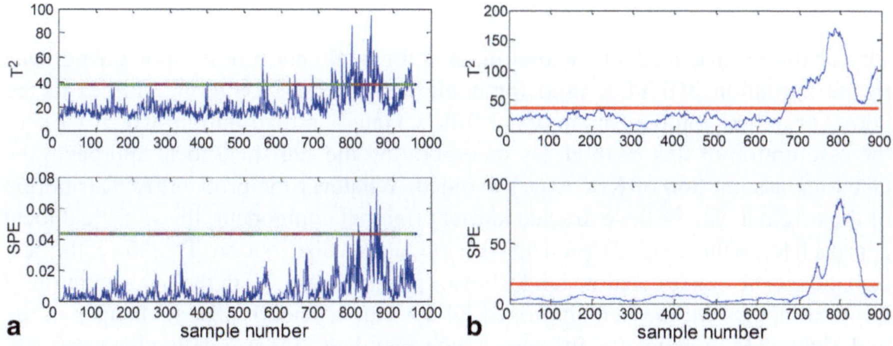

Fig. 5.9 Monitoring results of fault 15. **a** KPCA. **b** New method

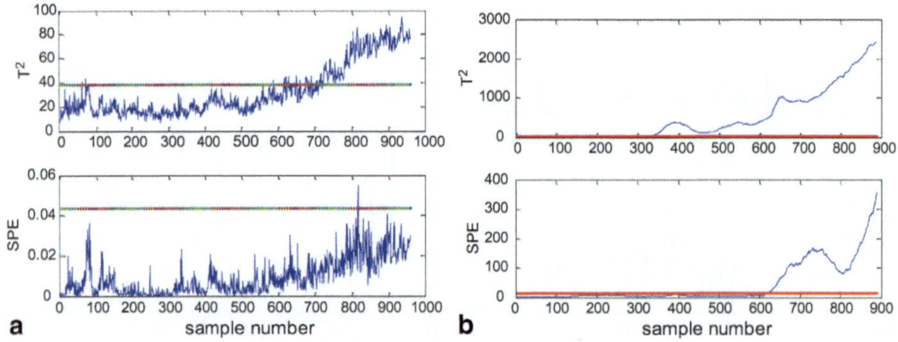

Fig. 5.10 Monitoring results of fault 21. **a** KPCA. **b** New method

fault. However, the conventional KPCA method can still hardly detect this fault, especially after sample 750. The improvement of the new method for this fault is also significant, which can be seen from Fig. 5.9b. The fault can be continuously detected in the late period of the process. The improvement of other small faults (i.e., faults 3 and 9) can be clearly found in Table 5.1. An unknown fault (fault 21) detection results are plotted in Fig. 5.10. The fault is detected at about sample 620 by KPCA, and it cannot be continuously detected until sample 720. However, the performance is greatly improved by the new method. The fault is continuously detected from sample 330 in Fig. 5.10b. With the introduction of the moving window in the new approach, the detection point is actually at sample 400, since the length of the moving window is 70. Therefore, this fault can be detected much earlier (320 sample time) by the new method. Again, the change value of the statistic in the new method is more significant than that in KPCA, which shows the new method is more sensitive than KPCA.

5.5 Conclusions

This chapter introduced a new joint method for nonlinear process monitoring. Due to the limitation of KPCA used for nonlinear process monitoring, the extracted kernel principal components may not follow Gaussian distribution, which violates the assumption of this method. By incorporating the statistical local approach, the Gaussian assumption of KPCA was avoided. Whatever the probability distribution of the original data or the extracted kernel principal component, the statistical local approach leads the method to end up with a Gaussian distribution. Therefore, the new statistic can be constructed which follows χ^2 distribution. With the incorporation of the local approach, the new improved KPCA can efficiently detect changes in the underlying KPCA geometry. In contrast, however, KPCA may not function very well. While the traditional monitoring statistics cannot detect incipient fault conditions, the improved residuals which represent geometric changes of the KPCA decomposition do not suffer from this inherent limitation. Therefore, based upon the two case studies, the new method seems to offer a more sensitive process monitoring capability than the existing work. However, there are still some problems to be solved to improve this method. First, the number of the new score variable is selected upon the principal component number. How to choose it optimally is important for process monitoring. Second, it is difficult to identify the root cause of the detect fault due to the nonlinear relationship between the original data and score variables. Besides, the computation burden becomes very large with the increase of the training samples.

5.6 Appendix: Proof of Theorem 5.1

As mentioned in Sect. 5.2.2, due to the simplicity and computation efficiency of the monitoring function Λ, it is employed for process monitoring. However, the monitoring performance between monitoring functions Λ and $\mathbf{\Omega}$ is equivalent, which was also highlighted by Kruger et al. (2007). Rewriting Eq. (5.5) as follows

$$\begin{aligned}\mathbf{v}_i &\triangleq \arg\max_{\mathbf{v}}\{\mathbf{E}(\mathbf{v}^T\bar{\Phi}(\mathbf{x})\bar{\Phi}(\mathbf{x})^T\mathbf{v}) - \lambda_i(\mathbf{v}^T\mathbf{v}-1)\} \\ &= \arg\max_{\alpha}\{\mathbf{E}(\alpha^T\bar{\mathbf{K}}(\mathbf{x})\bar{\mathbf{K}}(\mathbf{x})^T\alpha) - \lambda_i(\alpha^T\bar{\mathbf{K}}\alpha-1)\}\end{aligned} \tag{5.27}$$

To investigate the impact of change in the kernel eigenvalue λ_i, denote the departure of λ_i as $\Delta\lambda_i$, thus

$$\lambda_i^{\#} = \lambda_i + \Delta\lambda_i \tag{5.28}$$

where $\lambda_i^{\#}$ is the kernel eigenvalue after its change. Notice Eq. (5.7) and $\mathbf{t}_i = \alpha_i\bar{\mathbf{K}}(\mathbf{x})$, the following equation arises

$$2\alpha_i\bar{\mathbf{K}}^2(\mathbf{x}) - 2\lambda_i\bar{\mathbf{K}}(\mathbf{x})\alpha_i = 0 \tag{5.29}$$

When the process (eigenvalue) changes, the kernel matrix $\bar{\mathbf{K}}(\mathbf{x})$ also changes, which can be denoted as

$$\bar{\mathbf{K}}^{\#}(\mathbf{x}) = \bar{\mathbf{K}}(\mathbf{x}) + \Delta\bar{\mathbf{K}}(\mathbf{x}) \tag{5.30}$$

Let $\mathbf{J} = \bar{\mathbf{K}}^2(\mathbf{x})$ and

$$\mathbf{J}^{\#} = \mathbf{J} + \Delta\mathbf{J} \tag{5.31}$$

Thus, Eq. (5.29) becomes as follows by multiplying α_i^T in each side and incorporating Eq. (5.31)

$$2\alpha_i^T\mathbf{J}^{\#}\alpha_i - 2\lambda_i\alpha_i^T\bar{\mathbf{K}}^{\#}(\mathbf{x})\alpha_i = \Xi \neq 0 \tag{5.32}$$

Notice $\bar{\mathbf{K}}^{\#}(\mathbf{x})\alpha_i = \lambda_i^{\#}\alpha_i, \mathbf{J}^{\#} = \bar{\mathbf{K}}^{\#2}(\mathbf{x}), \alpha_i^T\bar{\mathbf{K}}^{\#}(\mathbf{x})\alpha_i$ and Eqs. (5.28) and (5.32) become as follows

$$\begin{aligned}\Xi &= 2\alpha_i^T\mathbf{J}^{\#}\alpha_i - 2\lambda_i\alpha_i^T\bar{\mathbf{K}}^{\#}(\mathbf{x})\alpha_i \\ &= 2\alpha_i^T\mathbf{J}^{\#}\alpha_i - 2\lambda_i^{\#}\alpha_i^T\bar{\mathbf{K}}^{\#}(\mathbf{x})\alpha_i + 2\Delta\lambda_i\alpha_i^T\bar{\mathbf{K}}^{\#}(\mathbf{x})\alpha_i \\ &= 2\Delta\lambda_i \end{aligned} \tag{5.33}$$

Therefore, if the kernel eigenvalue changes, the value of the monitoring function Λ departs from zero (both in systematic and noisy parts). This implies that both primary residuals of $\Lambda_s$and Λ_n are sufficient in detecting changes in kernel eigenvalues.

Chapter 6
Nonlinear Process Monitoring: Part 2

6.1 Introduction

Along last several decades, successful studies of data-based process monitoring methods such as multivariate statistical process control (MSPC) have been reported both in academic and industrial areas (Kruger et al. 2001; Qin 2003; Wang et al. 2004; Simoglou et al. 2005; Kruger and Dimitriadis 2008; Zhao et al. 2007). As one of the most popular MSPC methods used for process monitoring, principal component analysis (PCA) has become very popular. However, the traditional PCA-based monitoring method has assumed that process variables are linear, Gaussian distributed, and operated under single mode. In reality, these restrictions can be easily violated. In other words, data obtained from complex processes are not always linear, Gaussian distributed, and they may come from different operation conditions. Compared to the linear process, it is more difficult to detect the fault in nonlinear processes. Process monitoring for nonlinear processes has become an important aspect in this area.

To address the nonlinear behavior of the process, several nonlinear extensions of the traditional PCA method have already been proposed. Kramer (1992) developed a nonlinear PCA method based on the autoassociative neural network. Dong and McAvoy (1996) proposed a nonlinear PCA by combining the principal curve and the neural network. Kruger et al. (2005) proposed a new nonlinearity measure for principal component and discussed whether the nonlinear PCA should be used. Recently, a new nonlinear process monitoring method based on kernel PCA (KPCA) has been proposed (Lee et al. 2004). The main idea of KPCA is first to map the input space into a feature space via a nonlinear map, which makes data structure more linear, and then to extract principal components in the feature space. Other related nonlinear PCA methods have also been reported, including Cheng and Chiu (2005); Maulud et al. (2006); Wang et al. (2007); Zhang and Qin (2008), etc. However, Most of them are computationally inefficient and difficult to be used, especially for online monitoring.

The motivation of this chapter is to introduce an improved nonlinear process monitoring method, which is also more computationally efficient. Different from traditional nonlinear methods, the introduced method intends to build multiple linear

Z. Ge, Z. Song, *Multivariate Statistical Process Control,* Advances in Industrial Control,
DOI 10.1007/978-1-4471-4513-4_6,

monitoring models for nonlinear processes. Precisely, the original nonlinear space is approximated by several linear subspaces, based on which different subspace models can be constructed. In this method, the diversity and the accuracy of each linear subspace are both important. As the principal directions of PCA are orthogonal to each other, if we build individual sub-models in these orthogonal subspaces, the diversity of linear subspaces will be obtained. On the other hand, if the most relevant variables are selected in each subspace for modeling, the accuracy of the individual sub-model can also be gained. Therefore, to get a balance between the accuracy and diversity of the subspace model, PCA decomposition is carried out for linear subspace construction. To obtain the final decision, monitoring results in different linear subspaces should be combined together. However, because each subspace has its own monitoring statistic and confidence limit, it is difficult to combine them directly. Here, the Bayesian inference method is used to transfer the traditional monitoring statistic value to a fault probability in each individual subspace (Bishop 2006; Ge et al. 2010b), hence, results in different subspaces can be easily combined together.

Based on the linear subsapce monitoring framework, a corresponding fault diagnosis method can be developed. As a traditional method, the contribution plot method and its variants have been used in many research works (Miller et al. 1993; Westerhuis et al. 2000). However, the contribution plot method may cause misleading results, because it has no rigorous analysis of diagnosability, and correct diagnosis results cannot be guaranteed. To this end, a geometric method has been developed for the analyses of fault detectability, reconstructability, and isolatability (Qin 2003; Dunia and Qin 1998a; Lieftuche et al. 2006b). More recently, a reconstruction-based contribution plot method (RBC) has been developed (Alcala and Qin 2009), which combined the reconstruction and contribution plot methods together, thus the correct diagnosis result can be guaranteed. Based on the theory foundation of RBC and its efficiency, it is employed for fault diagnosis in the current monitoring framework. Precisely, RBC is used for fault diagnosis in each linear subspace, and then these subspace results are combined together to form the final diagnosis decision.

6.2 Linear Subspace Development

Suppose the collected process dataset is denoted as $\mathbf{X} \in R^{n \times m}$, where m is the number of process variables, and n is the sample number for each variable. The original nonlinear space can be denoted as

$$\mathbf{X} = \Phi(\mathbf{T}_{nl}) + \mathbf{E}_{nl} \tag{6.1}$$

where $\Phi(\cdot)$ defines the nonlinear relationship between the process variables, $\mathbf{T}_{nl}$ is the latent variable matrix, and $\mathbf{E}_{nl}$ is the residual matrix. As mentioned above, the nonlinear process can be approximated by several linear subspaces. Clearly, the linear subspace method can be described in Fig. 6.1, in which the nonlinear curve

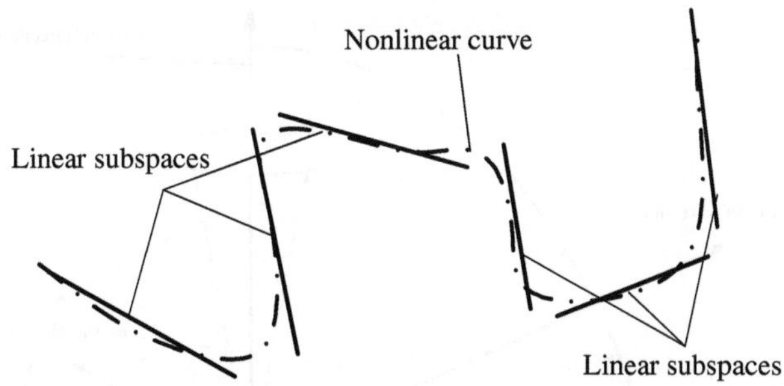

Fig. 6.1 Description of the linear subspace method

is denoted as the nonlinear relationship between different variables. To approximate the nonlinear curve, several linear subspaces could be used, which are also highlighted in Fig. 6.1. An important issue of this method is how to guarantee both diversity and accuracy of the individual subspace. Here, these linear subspaces are constructed by PCA decomposition. Therefore, the traditional PCA decomposition is firstly implemented on the dataset **X**, thus

$$\mathbf{X} = \mathbf{T}\mathbf{P}^T + \tilde{\mathbf{T}}\tilde{\mathbf{P}}^T \tag{6.2}$$

where the first item represents principal components extracted in the principal component subspace (PCS), and the second item corresponds to the residual subspace (RS). If k principal components are selected in PCS, then the loading matrices in PCS and RS can be decomposed as follows

$$\mathbf{P} = [\mathbf{P}_1, \mathbf{P}_2, \ldots, \mathbf{P}_k] \tag{6.3}$$

$$\tilde{\mathbf{P}} = [\tilde{\mathbf{P}}_1, \tilde{\mathbf{P}}_2, \ldots, \tilde{\mathbf{P}}_{m-k}] \tag{6.4}$$

According to the PCA method, the items in Eqs. (6.3) and (6.4) are orthogonal with each other. It is noticed that if we build individual sub-models in these orthogonal subspaces, the diversity of linear subspaces will be obtained. On the other hand, if the most relevant variables are selected in each subspace for modeling, the accuracy of the individual sub-model can also be gained. The sketch of the PCA-based linear subspace construction method is described in Fig. 6.2. As can be seen, for the construction of the ith subspace, those variables which are most correlated with the ith principal component are selected in this subspace. Hence, the diversity and accuracy of the method can both be guaranteed (Ge et al. 2010).

Therefore, the nonlinear space can be approximated by $k+1$ linear subspaces as follows (Ge et al. 2010)

$$subspace\{i,\ i = 1, 2, \ldots, k\} = \mathbf{P}\{i,\ i = 1, 2, \ldots, k\} \tag{6.5}$$

$$subspace\{k+1\} = mean\ (\tilde{\mathbf{P}}\{i,\ i = 1, 2, \ldots, m-k\}) \tag{6.6}$$

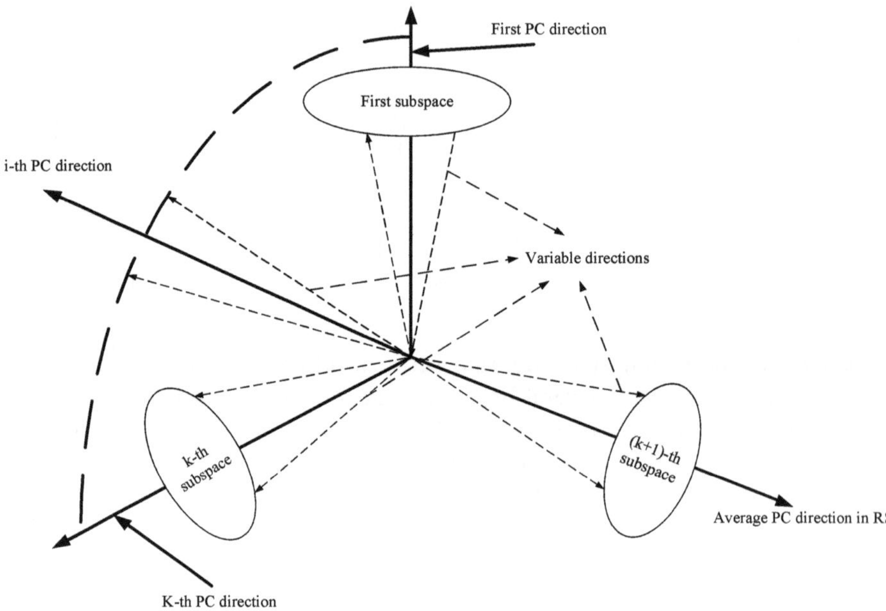

Fig. 6.2 Sketch of the subspace construction method

Before developing individual sub-models in these linear subspaces, a subset of process variables should be selected in each subspace. Here, we define a subspace contribution index to measure the importance of the variable in each subspace. As we have constructed linear subspaces in both PCS and RS, two different kinds of subspace contribution indices should be defined, which are given as follows

Definition 6.1 The contribution of variable i in subspace j constructed in PCS can be calculated as

$$SubCI_{PCS}(i,j) = \frac{|p_{ij}|}{|p_{1j}| + \cdots + |p_{ij}| + \cdots + |p_{mj}|} \tag{6.7}$$

where $i = 1, 2, \ldots, m$, $j = 1, 2, \ldots, k$, and p_{ij} is the ith row jth column element of the principal loading matrix $\mathbf{P}$.

Definition 6.2 The contribution of variable i in the subspace constructed in RS can be calculated as

$$SubCI_{RS} = \frac{|q_i|}{|q_1| + \cdots + |q_i| + \cdots + |q_m|}$$
$$|q_i| = mean\left([\,|\tilde{p}_{i1}|\,|\tilde{p}_{i2}| \cdots |\tilde{p}_{i(m-k)}|\,]\right) \tag{6.8}$$

where $i = 1, 2, \ldots, m$, $\tilde{p}_{ij}$ is the ith row jth column element of the residual loading matrix $\tilde{\mathbf{P}}$, and $mean(\cdot)$ represents the mean value of the elements.

Therefore, to select the variable index for the subspace j constructed in PCS, those which have big contribution values $SubCI_{PCS}(i, j), i = 1, 2, \ldots, m$ for subspace j are determined as the variable subset. Suppose the size of the subspace is selected as m_j, variable indices which correspond to the highest m_j contribution values should be determined as the index subset for the subspace j. Similarly, the index subset for the subspace constructed in RS can be determined through the contribution value $SubCI_{RS}$. When the variable index subset has been selected in each linear subspace, we are in the position to develop individual sub-models. Denote the subspace datasets as (in MATLAB manner)

$$\mathbf{X}_i = \mathbf{X}(:, \mathbf{Ind}(i)), \quad i = 1, 2, \ldots, k + 1 \tag{6.9}$$

where $\mathbf{Ind}(i)$ is the selected variable index for subspace i. Then, the individual PCA sub-model can be developed in each subspace as

$$\mathbf{X}_i = \mathbf{T}_i \mathbf{P}_i^T + \mathbf{E}_i, \quad i = 1, 2, \ldots, k + 1 \tag{6.10}$$

The number of principal components in each subspace can also be determined by the CPV rule or other cross-validation methods. Therefore, a total of $k + 1$ PCA sub-models can be developed.

6.3 Nonlinear Process Monitoring Based on Linear Subspace Method

6.3.1 Process Monitoring Approach

When all the linear subspaces have been constructed, the traditional monitoring statistics T^2 and SPE can be developed in each subspace as follows

$$T_i^2 = \sum_{j=1}^{k_i} \frac{\mathbf{t}_j^2}{\lambda_j} \le \frac{k_i(n-1)}{n-k_i} F_{k_i,(n-k_i),\alpha} = T_{i,\lim}^2 \tag{6.11}$$

$$SPE_i = \mathbf{e}\mathbf{e}^T = \mathbf{x}(\mathbf{I} - \mathbf{P}_i \mathbf{P}_i^T)\mathbf{x}^T \le SPE_{i,\lim} \tag{6.12}$$

$$\begin{aligned} SPE_{i,\lim} &= g_i \chi_{h_i,\alpha}^2 \\ g_i &= v_i/(2m_i) \\ h_i &= 2m_i^2/v_i \end{aligned} \tag{6.13}$$

where k_i is the number of PCs in the ith subspace, α is significance level, c_α is the normal deviate corresponding to the upper $1 - \alpha$ percentile, m_i and v_i are the mean and variance values of SPE within the ith subspace, $T_{i,\lim}^2$ and $SPE_{i,\lim}$ correspond to the confidence limits of T^2 and SPE statistics in the ith subspace, respectively.

Generally, each individual PCA monitoring model has its own confidence limit, because the retained number of PCs in these subspaces may be different. Thus, for a new monitored data sample $\mathbf{x}_t$, it is difficult to combine its subspace results directly into the final nonlinear monitoring decision. In our opinion, it is an efficient way to combine the subspace monitoring results through the probabilistic manner. Therefore, the Bayesian inference strategy is employed to transfer the monitoring statistic value to the fault probability in each subspace. First, the fault probability corresponding to the T^2 statistic in each subspace can be calculated as follows (Bishop 2006).

$$P_{T^2}(F|\mathbf{x}_{t,i}) = \frac{P_{T^2}(\mathbf{x}_{t,i}|F)P_{T^2}(F)}{P_{T^2}(\mathbf{x}_{t,i})}$$
$$P_{T^2}(\mathbf{x}_{t,i}) = P_{T^2}(\mathbf{x}_{t,i}|N)P_{T^2}(N) + P_{T^2}(\mathbf{x}_{t,i}|F)P_{T^2}(F) \tag{6.14}$$

where "N" and "F" represent normal and abnormal, respectively. The $\mathbf{x}_{t,i}$ $(i = 1,2,\ldots,k+1)$ is the selected data vector in the ith subspace, $P_{T^2}(N)$ and $P_{T^2}(F)$ are the prior probabilities of the process being normal and abnormal. When the significance level is selected as α, the values of $P_{T^2}(N)$ and $P_{T^2}(F)$ can be simply determined as $1-\alpha$ and α. However, to calculate the fault probability $P_{T^2}(F|\mathbf{x}_{t,i})$, we should further obtain two conditional probabilities $P_{T^2}(\mathbf{x}_{t,i}|N)$ and $P_{T^2}(\mathbf{x}_{t,i}|F)$. Here, these two probabilities are defined as follows (Ge et al. 2010)

$$P_{T^2}(\mathbf{x}_{t,i}|N) = \exp\left\{-\frac{T_i^2(\mathbf{x}_{t,i})}{\nu T_{i,\lim}^2}\right\}, \quad P_{T^2}(\mathbf{x}_{t,i}|F) = \exp\left\{-\frac{T_{i,\lim}^2}{\nu T_i^2(\mathbf{x}_{t,i})}\right\} \tag{6.15}$$

where ν is a tuning parameter and

$$T_i^2(\mathbf{x}_{t,i}) = \sum_{j=1}^{k_i} \frac{\mathbf{t}_{i,j}^2}{\lambda_j}, \quad \mathbf{t}_i = \mathbf{x}_{t,i}\mathbf{P}_i, \ i = 1,2,\ldots,k+1 \tag{6.16}$$

Similarly, the fault probability corresponding to the *SPE* statistic in each subspace can be calculated as follows

$$P_{SPE}(F|\mathbf{x}_{t,i}) = \frac{P_{SPE}(\mathbf{x}_{t,i}|F)P_{SPE}(F)}{P_{SPE}(\mathbf{x}_{t,i})}$$
$$P_{SPE}(\mathbf{x}_{t,i}) = P_{SPE}(\mathbf{x}_{t,i}|N)P_{SPE}(N) + P_{SPE}(\mathbf{x}_{t,i}|F)P_{SPE}(F) \tag{6.17}$$

where the conditional probabilities $P_{SPE}(\mathbf{x}_{t,i}|N)$ and $P_{SPE}(\mathbf{x}_{t,i}|F)$ are defined as

$$P_{SPE}(\mathbf{x}_{t,i}|N) = \exp\left\{-\frac{SPE_i(\mathbf{x}_{t,i})}{\nu SPE_{i,\lim}}\right\}, \quad P_{SPE}(\mathbf{x}_{t,i}|F) = \exp\left\{-\frac{SPE_{i,\lim}}{\nu SPE_i(\mathbf{x}_{t,i})}\right\} \tag{6.18}$$

where

$$SPE_i(\mathbf{x}_{t,i}) = \mathbf{x}_{t,i}(\mathbf{I} - \mathbf{P}_i\mathbf{P}_i^{\mathrm{T}})\mathbf{x}_{t,i}^T, \quad i = 1,2,\ldots,k+1 \tag{6.19}$$

Then, these two calculated fault probabilities in different linear subspaces can be combined together to form into two final nonlinear monitoring statistics: $BIC_{T^2}(\mathbf{x}_t)$ and $BIC_{SPE}(\mathbf{x}_t)$. Basically, the final monitoring statistic values can be determined as any value between the minimum and maximum subspace values, thus

$$\min_{i=1,2,\ldots,k+1}\{P_{T^2}(F|\mathbf{x}_{t,i})\} \le BIC_{T^2}(\mathbf{x}_t) \le \max_{i=1,2,\ldots,k+1}\{P_{T^2}(F|\mathbf{x}_{t,i})\} \tag{6.20}$$

$$\min_{i=1,2,\ldots,k+1}\{P_{SPE}(F|\mathbf{x}_{t,i})\} \le BIC_{SPE}(\mathbf{x}_t) \le \max_{i=1,2,\ldots,k+1}\{P_{SPE}(F|\mathbf{x}_{t,i})\} \tag{6.21}$$

However, if the $BIC_{T^2}(\mathbf{x}_t)$ or $BIC_{SPE}(\mathbf{x}_t)$ is selected as one of the extreme values in Eqs. (6.20) and (6.21), the monitoring strategy may be too sensitive or too conservative, which are neither favorable for process monitoring. For simplicity, $BIC_{T^2}(\mathbf{x}_t)$ and $BIC_{SPE}(\mathbf{x}_t)$ can also be selected as the mean value among all subspaces. Alternatively, we utilize the weighted form to combine the subspace monitoring results, which are given as follows

$$BIC_{T^2}(\mathbf{x}_t) = \sum_{i}^{k+1}\left\{\frac{P_{T^2}(\mathbf{x}_{t,i}|F)P_{T^2}(F|\mathbf{x}_{t,i})}{\sum_{j}^{k+1} P_{T^2}(\mathbf{x}_{t,j}|F)}\right\} \tag{6.22}$$

$$BIC_{SPE}(\mathbf{x}_t) = \sum_{i}^{k+1}\left\{\frac{P_{SPE}(\mathbf{x}_{t,i}|F)P_{SPE}(F|\mathbf{x}_{t,i})}{\sum_{j}^{k+1} P_{SPE}(\mathbf{x}_{t,j}|F)}\right\} \tag{6.23}$$

When the value of $BIC_{T^2}(\mathbf{x}_t)$ or $BIC_{SPE}(\mathbf{x}_t)$ exceeds the significance level α, some process change or fault is considered to be detected. Otherwise, the process should be deemed as normal.

6.3.2 *Fault Diagnosis*

When some process fault has been detected, it should be correctly diagnosed, thus the root cause of the fault should be found and fixed up. To introduce the RBC method, the normal value of the monitored sample $\mathbf{x}_t$ in the ith subspace is assumed to be $\mathbf{x}^*_{t,i}$, the relationship between $\mathbf{x}^*_{t,i}$ and $\mathbf{x}_{t,i}$ can be represented as

$$\mathbf{x}^*_{t,i} = \mathbf{x}_{t,i} - \theta_q \mathbf{f}_{i,q} \tag{6.24}$$

where $\mathbf{f}_{i,q}$ is the estimated fault vector with $\|\mathbf{f}\|$ as its magnitude, and θ_q is the qth column of the identity vector. The reconstruction-based contribution of $\mathbf{x}_{t,i}$ along the qth variable direction can be calculated as follows (Alcala and Qin 2009)

$$RBC_{i,q} = \left\| \mathbf{M}^{1/2} \theta_q \mathbf{f}_q \right\|^2 = \mathbf{x}_{t,i}^T \mathbf{M} \theta_q (\theta_q^T \mathbf{M} \theta_q)^{-1} \theta_q^T \mathbf{M} \mathbf{x}_{t,i} = \frac{(\theta_q^T \mathbf{M} \mathbf{x}_{t,i})^2}{m_{qq}} \tag{6.25}$$

$$\mathbf{f}_{i,q} = (\theta_q^T \mathbf{M} \theta_q)^{-1} \theta_q^T \mathbf{M} \mathbf{x}_{t,i} \tag{6.26}$$

where $m_{qq} = \theta_q^T \mathbf{M} \theta_q$ is the qth diagonal element of $\mathbf{M}$. The matrix $\mathbf{M}$ is different for each monitoring statistic, which can be referred to Alcala and Qin (2009). Corresponding to the *SPE* statistic, $\mathbf{M}$ is equal to $\tilde{\mathbf{P}}_i \tilde{\mathbf{P}}_i^T$. However, in this part, we only present the RBC result based on the *SPE* statistic.

When all RBC values along different variable directions in each subspace have been calculated, the subspace diagnosis results can be combined as follows (Ge et al. 2010)

$$RBC_{gi} = [RBC_i \quad \mathbf{0}\{\mathbf{I\tilde{n}d}(i)\}] \tag{6.27}$$

$$\mathbf{f}_{gi} = [\mathbf{f}_i \quad \mathbf{0}\{\mathbf{I\tilde{n}d}(i)\}] \tag{6.28}$$

$$RBC_{comg,q} = \sum_i^{k+1} \left\{ \frac{P_{SPE}(\mathbf{x}_{t,i}|F) RBC_{gi,q}}{\sum_j^{k+1} P_{SPE}(\mathbf{x}_{t,j}|F)} \right\} \tag{6.29}$$

$$\mathbf{f}_{comg,q} = \sum_i^{k+1} \left\{ \frac{P_{SPE}(\mathbf{x}_{t,i}|F) \mathbf{f}_{gi,q}}{\sum_j^{k+1} P_{SPE}(\mathbf{x}_{t,j}|F)} \right\} \tag{6.30}$$

where $i = 1, 2, \ldots, k+1$, $q = 1, 2, \ldots, m$, RBC_{gi} and $\mathbf{f}_{gi}$ represent the global RBC and $\mathbf{f}$ values, which contains all the m process variables. The $\mathbf{0}\{\mathbf{I\tilde{n}d}(i)\}$ is a zero vector with variable indices which are not included in the ith subspace. The $RBC_{gi,q}$ and $\mathbf{f}_{gi,q}$ are qth elements of RBC_{gi} and $\mathbf{f}_{gi}$, respectively. The $RBC_{comg,q}$ and $\mathbf{f}_{comg,q}$ represent the final RBC and $\mathbf{f}$ values corresponding to the qth variable. Through analysis of Eq. (6.29), we can see that the probability of the monitored sample $\mathbf{x}_t$ in each subspace has been considered for fault diagnosis. If the fault happens in the ith subspace with high probability, then the contribution of that subspace should be put as a significant value through the combination step. Therefore, depending on the RBC and the linear subspace combination methods, the final diagnosis results can be obtained for the detected fault.

6.4 Method Implementation and Remarks

The implementation procedures of the new method can be summarized as follows

1. *Process monitoring*

Step 1 Collect a historical dataset that runs under normal process condition.
Step 2 Construct linear subspaces based on the PCA decomposition method and calculate the values of the subspace contribution index in each linear subspace.
Step 3 Develop subspace PCA models for the constructed subspaces, determine the confidence limits of their corresponding monitoring statistics.
Step 4 For each monitored sample, calculate its posterior probability in each subspace based on Bayesian inference by Eqs. (6.14) and (6.17).
Step 5 Combine the results obtained in all subspaces through Eqs. (6.22) and (6.23).
Step 6 Generate the new monitoring chart and detect process abnormalities for the monitored samples which satisfy $BIC_{T^2} > \alpha$ or $BIC_{SPE} > \alpha$.

2. *Fault diagnosis*

Step 1 For a detected faulty sample, the RBC values of process variables in each subspace and the corresponding magnitude values are calculated first through Eqs. (6.25) and (6.26).
Step 2 Extend the calculated results in each subspace to the global version by Eqs. (6.27) and (6.28).
Step 3 Combine the diagnosis results of each subspace through Eqs. (6.29) and (6.30).
Step 4 Make the final decision to determine the root cause of the detected fault.

Compared to the conventional monitoring method, which only incorporates a single model, the new method builds multiple models in different linear subspaces. Both accuracy and diversity of the constructed subspaces can be guaranteed by the PCA decomposition method. Due to the difficulty of result combination in different subspaces, the two traditional monitoring statistics are transferred to fault probabilities by Bayesian inference. For fault diagnosis, the RBC method is employed and similar combination step is taken to form the final diagnosis result. It is noticed that this method is a linear combination method, since all subspace monitoring models are based on the linear PCA method. Hence, the computational complexity of the new method has the same quantity as linear PCA. In the next section, we can find that the monitoring performance can be greatly improved by the new method while its computation complexity will not grow exponentially. The detailed analyses of algorithm complexities for PCA, KPCA, and the new method are given in the Appendix. For illustration simplicity, we denote the new method as BSPCA.

6.5 Case Studies

In this section, two case studies are provided to evaluate the performance of the BSPCA method. The first one is a numerical example with five variables, and the other one is the TE benchmark process, which has been widely used for process modeling, monitoring, and control algorithm evaluations in the past decades (Singhai and Seborg 2006; Ge et al. 2009; Kano et al. 2002; Chiang et al. 2001).

6.5.1 Numerical Example

The five-variable nonlinear system is constructed by the following equation

$$
\begin{aligned}
x_1 &= y + e_1 \\
x_2 &= y^2 - 3y + e_2 \\
x_3 &= -y^3 + 3y^2 + e_3 \\
x_4 &= y^4 - 4y^3 + 2y + e_4 \\
x_5 &= -2y^5 + 6y^4 - 3y^3 + y + e_5
\end{aligned} \tag{6.31}
$$

where y follows uniform distribution between [0.01, 2], e_1, e_2, e_3, e_4, and e_5 are independent Gaussian noise variables distributed as with zero mean and variance 0.01.

For PCA model construction, 200 data samples were generated under the normal process condition, and scaled to zero mean and variance 1, which means the mean and standard deviation values have both been removed from the data. To determine the number of retained principal components, the cumulative percentage variance (CPV) rule is used, by which the number was selected when $CPV > 85\ \%$. Hence, one principal component was selected in this system, which can explain 90.65 % of process information. Therefore, two subspaces were constructed. In this example, three variables were selected in each subspace, which were determined by the subspace contribution index. Then, two PCA sub-models are developed in the two linear subspaces.

In order to test the performance of the BSPCA method, three additional test datasets were generated, including one normal and two fault datasets, which all consist of 200 samples. The two fault cases are generated as follows:

Case 1 A step change of x_3 by 1 was introduced starting from sample 101.
Case 2 A ramp change of x_5 from sample 101 – 200 by adding 0.05 $(w - 100)$ to each sample, where w is the sample number.

First, the normal dataset was simulated. The monitoring results of BSPCA and the traditional PCA method are given in Fig. 6.3a, 6.3b. All dashed lines represent 99 % confidence limits of the corresponding statistics. As can be seen, the monitoring results of both methods indicate that the process is in good condition, since none of the monitoring statistics exceeds the confidence limit. It can be inferred that the BSPCA method does not lose efficiency in monitoring normal process data. To examine the monitoring situations in the two linear subspaces, monitoring results of

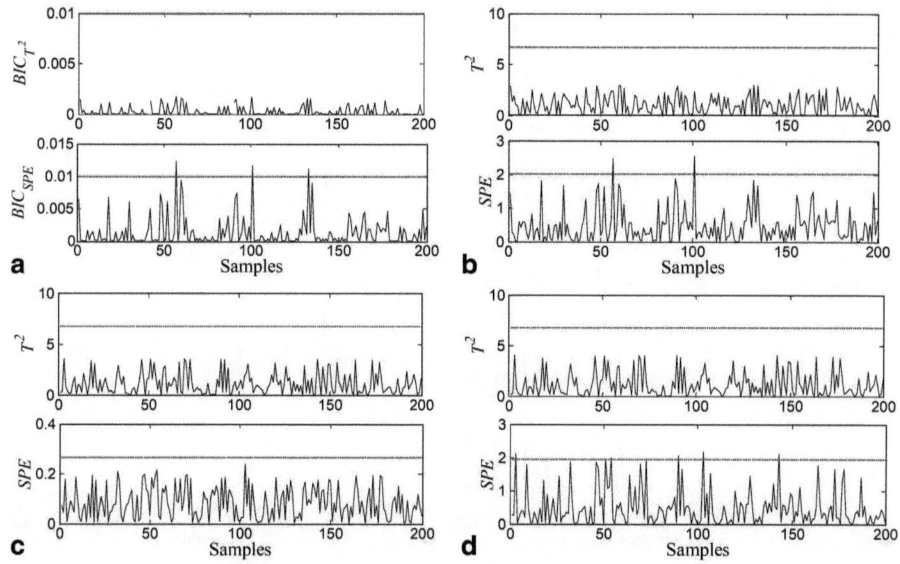

Fig. 6.3 Process monitoring results of normal process. **a** BSPCA. **b** PCA. **c** First subspace. **d** Second subspace

the two PCA sub-models are given in Fig. 6.3c, 6.3d. As expected, both monitoring results indicate that the process is running under normal operation condition.

Next, the two fault cases are tested. The monitoring results of the first case are given in Fig. 6.4a, 6.4b, which is a step change of the third variable. As shown in Fig. 6.4b, the traditional PCA method cannot detect any process abnormality, since both the monitoring statistics are inside their corresponding confidence limits. In contrast, as seen from Fig. 6.4a, the BSPCA method can successfully detect the fault, since the BIC_{SPE} statistic exceeds its confidence limit after sample 100. Detailed monitoring results in the two linear subspaces of this fault are given in Fig. 6.4c, 6.4d, in which we can find that the fault is mainly detected in the first subspace. This is because the selected variables for constructing the first subspace include variable 1, 3, and 5, while the second subspace consists of variable 2, 4, and 5. Therefore, when the fault happens in variable 3, the first subspace can successfully detect it. Similarly, the monitoring results of the second fault are shown in Fig. 6.5a, 6.5b, with the two subspace results given in Fig. 6.5c, 6.5d. Different from the first fault case, this fault is a ramp change of variable 5. Therefore, a ramp trend of the BIC_{SPE} monitoring statistic can be seen in Fig. 6.5a. In constrast, the tradition PCA method can hardly detect this fault.

After the fault has been detected, the efficiency of the new fault diagnosis method can be evaluated. First, one fault diagnosis result of the combined RBC method is given in Fig. 6.6a. It is very clear that the right cause (variable 3) of the fault has been identified. To examine the fault diagnosis results in each subspace, the results

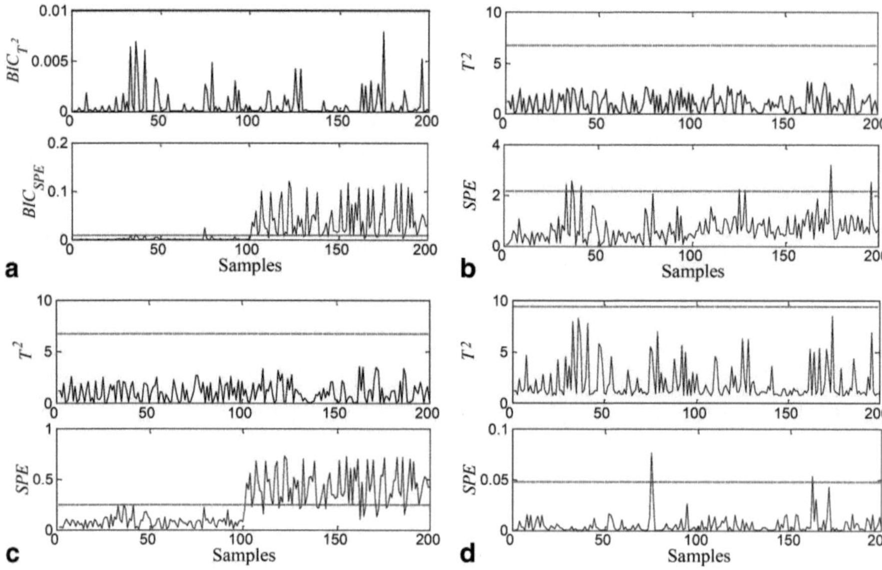

Fig. 6.4 Process monitoring results of fault case 1. **a** BSPCA. **b** PCA. **c** First subspace. **d** Second subspace

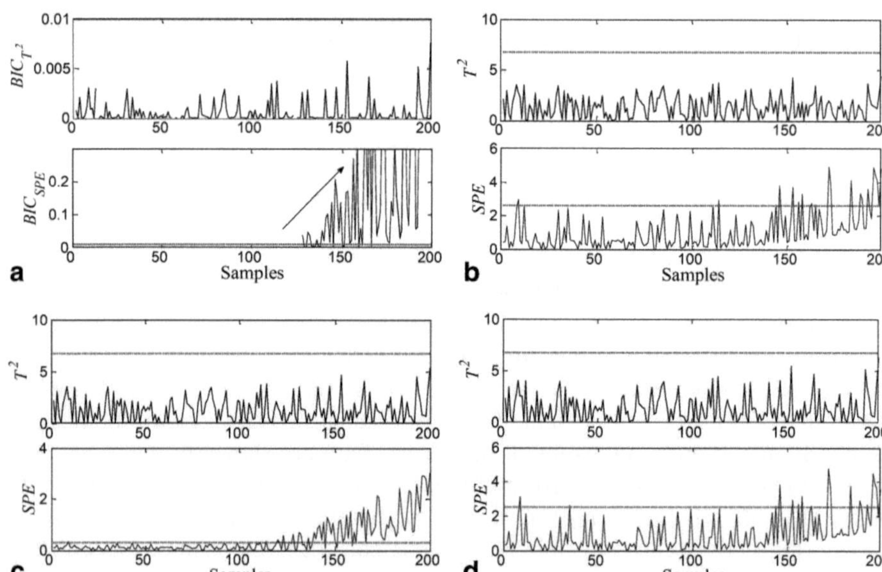

Fig. 6.5 Process monitoring results of fault case 2. **a** BSPCA. **b** PCA. **c** First subspace. **d** Second subspace

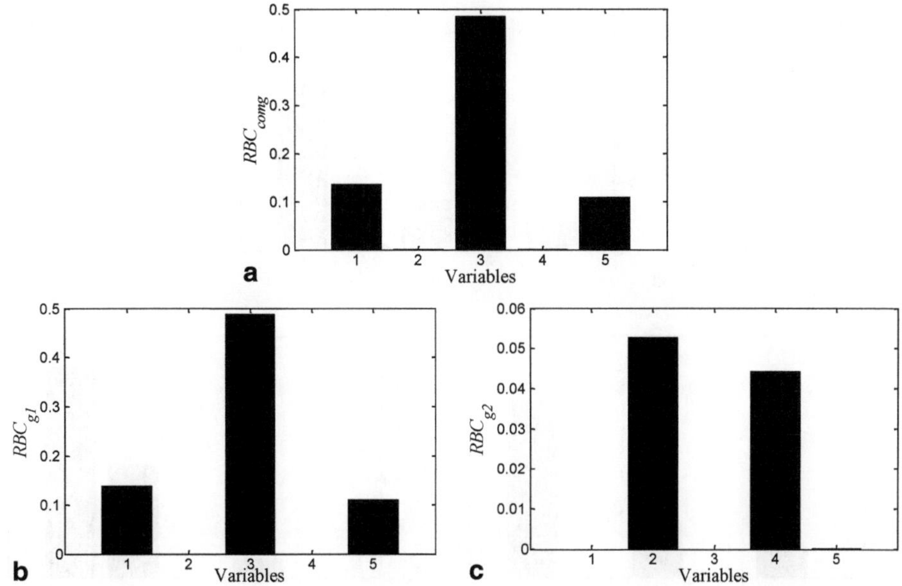

Fig. 6.6 Fault diagnosis results for fault case 1. **a** Combined RBC. **b** First subspace. **c** Second subspace

are given in Fig. 6.7b, 6.7c. As variable 3 was selected in the first subspace, it can be correctly diagnosed in the corresponding subspace. Comparing the results of Fig. 6.6a and 6.6b, one can find that they are very similar. This is because the fault probability in the first subspace is high while the value is very small in the second subspace. Therefore, the final fault diagnosis result is inclined to the first subspace. Similarly, fault diagnosis results of the second fault case are given in Fig. 6.7a, with its subspace diagnosis results presented in Fig. 6.7b, 6.7c. Besides, the magnitude of these two faults can also be estimated by the new method. The estimated results of all fault samples in the two fault cases are given in Figs. 6.8 and 6.9. As can be seen in Fig. 6.8, the estimated values of the first fault corresponding to the direction of variable 3 are around 1, which is the true value of the fault. The estimated values for the second fault are a little biased, which can be found in Fig. 6.9. However, the closest values to the real one are obtained through the direction of variable 5, which is the true fault direction of this fault. Besides, a ramp-type trend of the estimated values is also obtained, which should be the correct behavior of the fault.

6.5.2 *TE Benchmark Process*

TE process is used as the testing example in this section. The detailed description of this process can be found in Chap. 3. For process monitoring purpose, 16 vari-

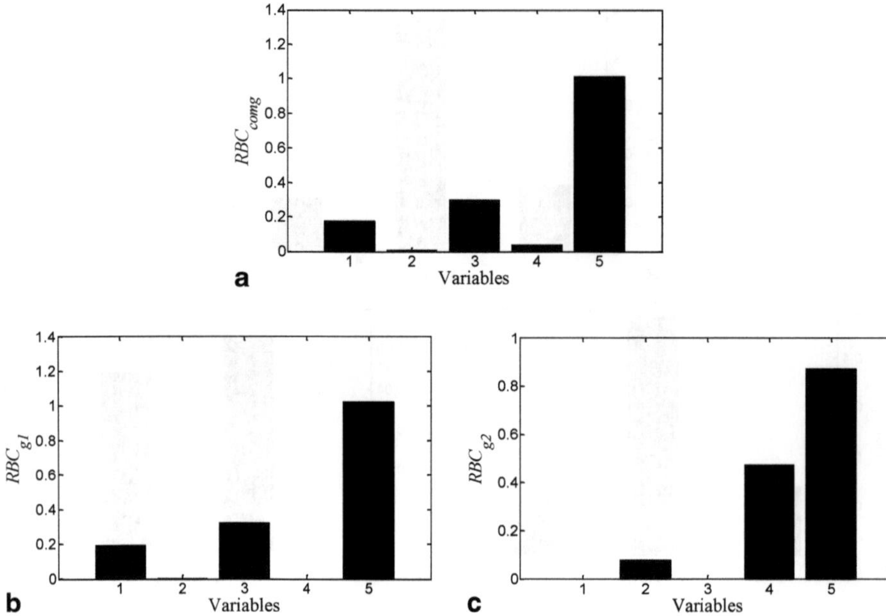

Fig. 6.7 Fault diagnosis results for fault case 2. **a** Combined RBC. **b** First subspace. **c** Second subspace

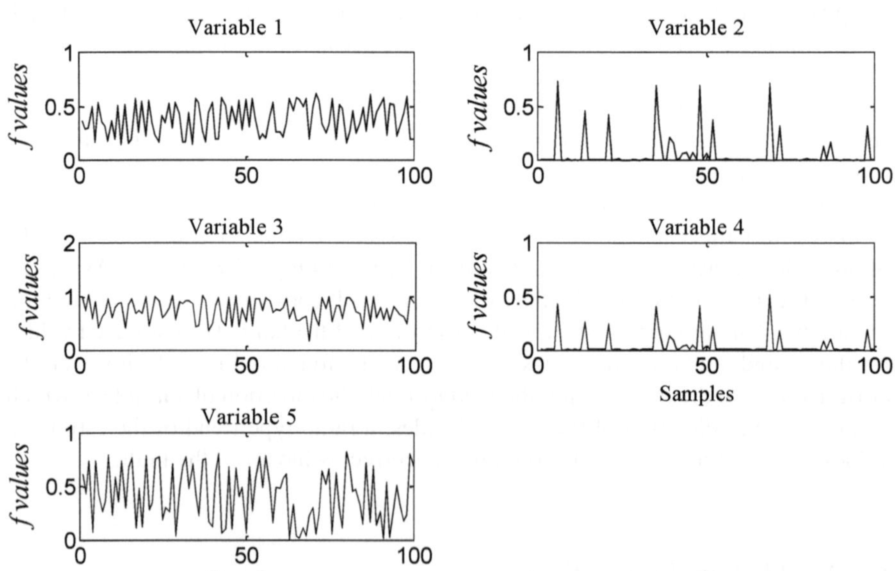

Fig. 6.8 Estimated fault magnitude of fault case 1

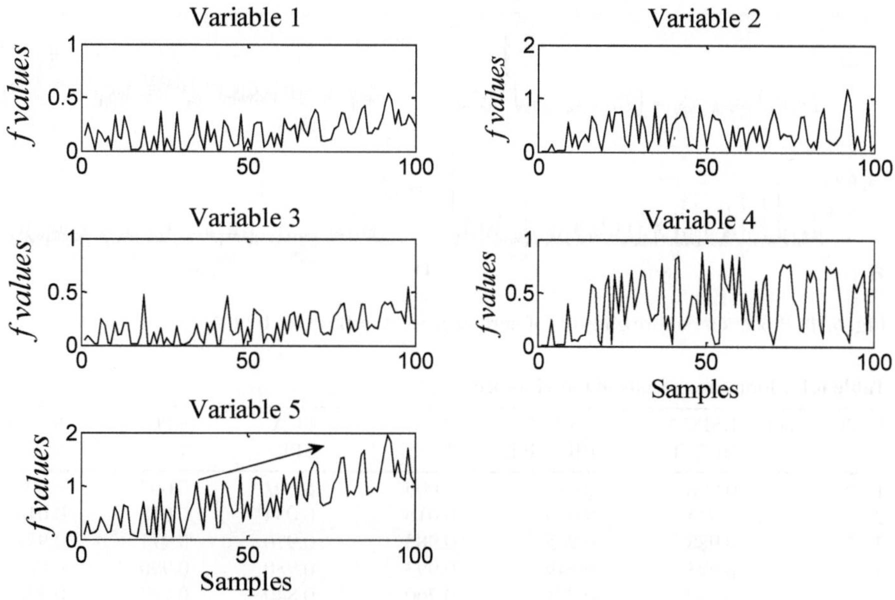

Fig. 6.9 Estimated fault magnitude of fault case 2

ables were selected, which is the same as that in Chaps. 4 and 5. A total of 960 data samples were generated for BSPCA model development. For comparison, the traditional PCA and KPCA models are also constructed. For performance evaluation, 22 testing datasets were also generated, which contain one normal scenario and 21 fault scenarios. Each testing dataset consists of 960 samples. All faults in the testing datasets were introduced in the process at sample 161. Faults 1–7 are step changes of process variables, faults 8 – 12 are random changes of variables, fault 13 is slow shift of Reaction kinetics, faults 14, 15, and 21 are related to valve sticking, and faults 16 – 20 are types of unknown faults. The detailed descriptions of the 21 faults can be found in Downs and Vogel (1993).

Depending on the CPV rule, the number of principal components for the PCA model is selected as 10, which can explain 87.75 % of the entire process information. Therefore, a total of 11 subspaces should be constructed. Eight process variables are selected in each subspace depending on their contributions. To evaluate the performance of the BSPCA method, 1 normal and 21 faulty testing datasets are used. First, the normal process monitoring results of BSPCA and PCA are given in Fig. 6.10. Both the monitoring results indicate that the process is operated under normal condition. After testing all the 21 process faults, the type II errors are tabulated together in Table 6.1 (mean values of 50 independent simulation times). Among this table, the minimum value of the type II error achieved for each fault is marked with a bold number. As shown in Table 6.1, BSPCA outperforms the conventional PCA and KPCA methods in most fault cases.

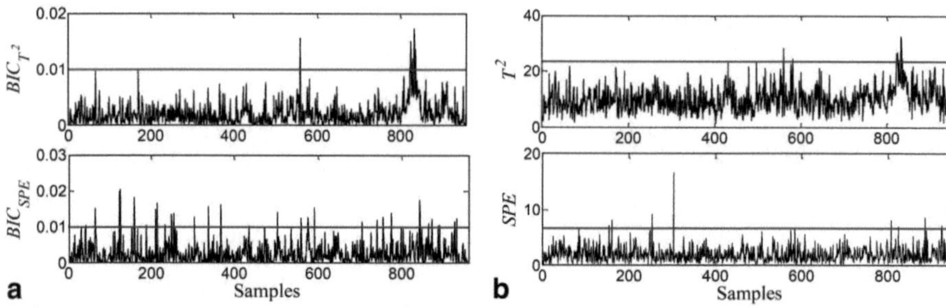

Fig. 6.10 Process monitoring results of normal process. **a** BSPCA. **b** PCA

Table 6.1 Monitoring results of the TE process

Fault number	BSPCA BIC_ T^2	BSPCA BIC_ SPE	PCA T^2	PCA SPE	KPCA T^2	KPCA SPE
1	0.008	*0.001*	0.009	*0.001*	0.003	0.004
2	*0.015*	*0.015*	0.016	0.034	0.018	0.018
3	0.988	*0.905*	0.983	0.976	0.971	0.979
4	0.993	*0.949*	0.993	0.980	0.990	0.988
5	0.769	*0.728*	0.760	0.840	0.763	0.756
6	*0.000*	*0.000*	*0.000*	0.003	*0.000*	*0.000*
7	0.620	0.609	0.625	0.751	*0.591*	0.595
8	0.029	0.026	0.028	0.166	*0.023*	*0.023*
9	0.980	*0.916*	0.980	0.975	0.975	0.970
10	0.659	0.433	0.691	0.660	0.608	*0.338*
11	0.570	*0.470*	0.860	0.484	0.575	0.609
12	0.011	0.026	0.016	0.115	*0.010*	0.013
13	0.058	*0.046*	0.059	0.075	0.050	0.055
14	*0.000*	*0.000*	0.001	*0.000*	*0.000*	0.001
15	0.970	*0.901*	0.974	0.973	0.923	0.938
16	0.750	*0.674*	0.765	0.894	0.703	0.686
17	0.110	*0.031*	0.178	0.033	0.040	0.064
18	0.106	*0.088*	0.106	0.094	0.099	0.103
19	0.850	0.880	*0.840*	0.934	0.881	0.864
20	0.728	0.509	0.784	0.615	0.693	*0.490*
21	0.611	*0.409*	0.626	0.578	0.574	0.584

Next, two process faults are selected to demonstrate the superiority of the BSPCA method. The first one is fault 3, which is a step change of D feed temperature. It is a very small process fault, thus can hardly be detected by the traditional PCA method. However, this fault can be partly detected by the BSPCA method. The monitoring results of both methods are shown in Fig. 6.11. The detected fault behaviors are highlighted by ellipses. To determine the root causes of this fault, the fault diagnosis results are given in Fig. 6.12a, which indicate that variable 8 and 14 take the most responsibilities of this fault. By examining the fault behavior in the process, it can be inferred that the purge rate and the stripper steam flow have strong relationship with the stream being changed. The estimated fault magnitudes through all the 16

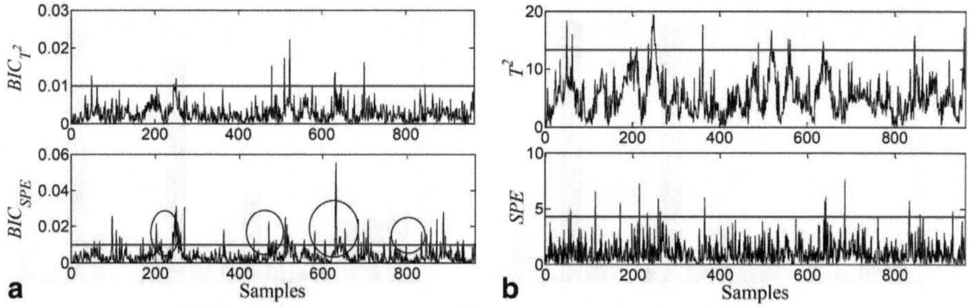

Fig. 6.11 Process monitoring results of fault 3. **a** BSPCA. **b** PCA

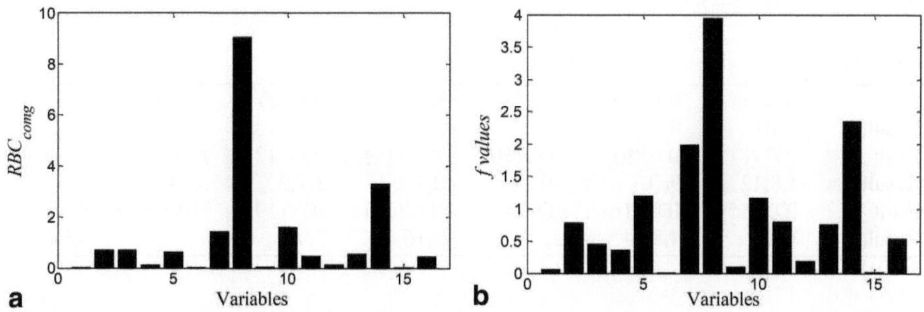

Fig. 6.12 Fault diagnosis and magnitude estimation results of fault 3. **a** Fault diagnosis. **b** Estimated fault magnitude

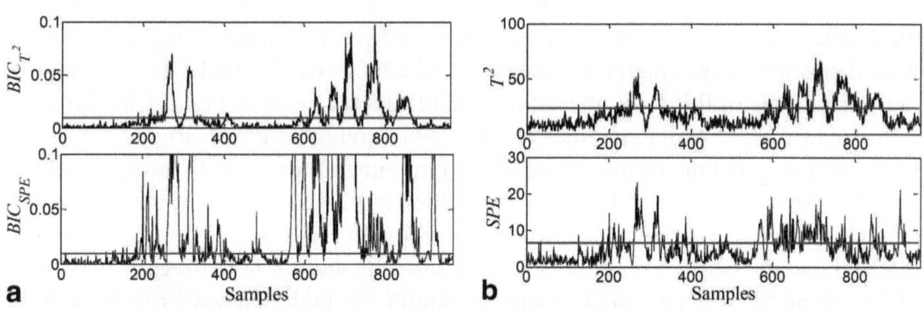

Fig. 6.13 Process monitoring results of fault 10. **a** BSPCA. **b** PCA

variable directions are shown in Fig. 6.12b. It can be seen that the biggest estimated fault magnitude is corresponding to variable 8, which is in accordance with the results given in Fig. 6.12a.

Similarly, the monitoring results of fault 10 based on both BSPCA and PCA methods are given in Fig. 6.13. It is a random change of the temperature of stream 4 (C feed) in the TE process. According to both results shown in Table 6.1 and Fig. 6.13, the monitoring performance of this fault has been greatly improved by the BIC_{SPE}

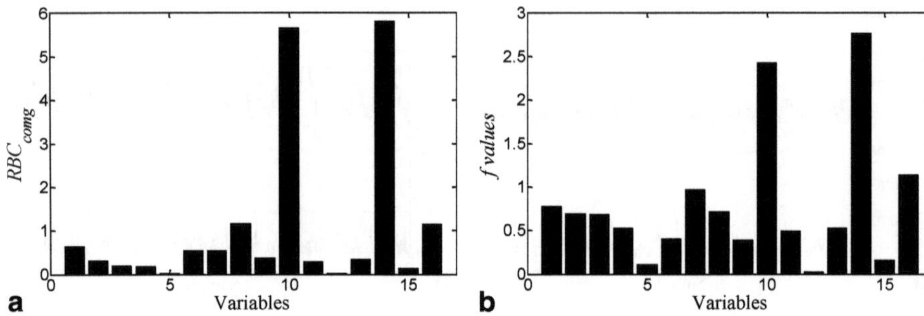

Fig. 6.14 Fault diagnosis and magnitude estimation results of fault 10. **a** Fault diagnosis. **b** Estimated fault magnitude

Table 6.2 Fault diagnosis results of all the 21 faults in TE process

Fault	IDV(1)	IDV(2)	IDV(3)	IDV(4)	IDV(5)	IDV(6)	IDV(7)
Results	1,10	8	8,14	2,8	3,12	3,9	3,11,14
Fault	IDV(8)	IDV(9)	IDV(10)	IDV(11)	IDV(12)	IDV(13)	IDV(14)
Results	1,8,12	9,16	10,14	2,7	3,7,9	3,14	7
Fault	IDV(15)	IDV(16)	IDV(17)	IDV(18)	IDV(19)	IDV(20)	IDV(21)
Results	8,12	4,8,14	2,15	9,16	7,9	10	3,14

statistic of the BSPCA method. The fault diagnosis results of this fault are given in Fig. 6.14a, which indicate that variable 10 and 14 are the most responsible for the fault. These two variables correspond to product separator pressure and stripper steam flow in the TE process. By examining and analyzing the flowchart of TE process, the product separator pressure and the stripper stream flow have strong relationships with the temperature change of stream 4, which causes the fault. To estimate the fault magnitude of this fault, the estimation results are given in Fig. 6.14b. Different estimated values of fault magnitude are obtained through different variable directions. It can be found that the estimated values of fault magnitude through the directions of variable 10 and 14 are much bigger than other estimated values. Again, the results obtained in Fig. 6.14b are in accordance with that in Fig. 6.14a. Therefore, it can be inferred that the BSPCA method is feasible and efficient for fault diagnosis. In order to determine the root causes of all process faults, the fault diagnosis results of these 21 faults are summarized in Table 6.2.

6.6 Conclusions

In this chapter, a nonlinear process monitoring method has been introduced, which was based on linear subspace and Bayesian inference. For linear subspace construction, the PCA decomposition method is used and two new subspace contribution indices have been defined. After the development of subspace monitoring models, the Bayesian inference strategy was utilized to transfer the traditional monitoring statistic to fault probability, which can be combined more easily. When some fault has been detected, a corresponding fault diagnosis approach has also been developed, which is based on the recently developed reconstruction-based method. Both the root cause and the estimated fault magnitude can be obtained simultaneously through the fault diagnosis step. Two case studies have been used to show the feasibility and efficiency of the new nonlinear process monitoring method.

6.7 Appendix: Algorithm Complexity Analyses of PCA, KPCA, and BSPCA

The algorithm complexity analyses of these three methods include offline modeling phase and online monitoring phase. First, through numerical analysis of the PCA method in the offline phase, it can be found that a large amount of time is spent in computing the covariance matrix for high-dimensional data. The time consumed in other steps of PCA is insignificant compared to the calculation of the covariance matrix (Kumar and Negi 2008). Therefore, we focus the offline analysis on time complexity of covariance matrices computation. The time complexity for calculating the $m \times m$ covariance matrix of the original dataset $\mathbf{X} \in R^{n \times m}$by PCA is given as follows

$$T_{PCA} = \begin{cases} O(nm^2) & \text{if } n > m \\ O(mn^2) & \text{otherwise} \end{cases} \tag{6.32}$$

where $n > m$ always holds in most processes. Now, we consider the time complexity of the constructed PCA sub-models in linear subspaces, suppose r variables have been selected in each subspace, Here, we assume that $n > r$ always holds, then the time complexity of the PCA sub-model can be calculated as

$$T_{SPCA} = O[(k+1)nr^2] \tag{6.33}$$

where $k+1$ is the number of linear subspaces. Compare Eq. (6.32) and Eq. (4.28), if $m > r\sqrt{k+1}$, then the time complexity of subspace PCA is lower than that of PCA, thus $T_{PCA} > T_{SPCA}$. Otherwise, if $m < r\sqrt{k+1}$, then $T_{PCA} < T_{SPCA}$ holds. However, it is straightforward that the time complexity of the BSPCA method is higher than either T_{PCA} or T_{SPCA}, because it is the summation of T_{PCA} and T_{SPCA}, thus

$$T_{BSPCA} = T_{PCA} + T_{SPCA} = O\{[(k+1)r^2 + m^2]n\} \tag{6.34}$$

In contrast, the algorithm complexity of KPCA is $O(n^3)$. Basically, $n \gg m$ is hold in most industrial processes, therefore, the following equation is also come into existence

$$O\{[(k+1)r^2 + m^2]n\} \ll O(n^3) \tag{6.35}$$

Next, the online algorithm complexity of the BSPCA method is analyzed. According to the online implementation of PCA, the complexity of this method can be calculated as

$$TO_{PCA} = mk + m + 2k \tag{6.36}$$

Then, the online complexity of the BSPCA method can be given as follows

$$TO_{BSPCA} = (k + 1)(rk_s + r + 2k_s) \tag{6.37}$$

where k_s is the selected number of principal components in each subspace, denote $\beta = m/k$, $\gamma = r/k_s$, then Eqs. (4.31) and (4.32) can be represented as

$$TO_{PCA} = O(\beta k^2) \tag{6.38}$$

$$TO_{BSPCA} = O(k\gamma k_s^2) \tag{6.39}$$

Comparing Eqs. (6.38) and (6.39), the following equation can be made

$$\tau = \frac{TO_{BSPCA}}{TO_{PCA}} = \frac{O(k\gamma k_s^2)}{O(\beta k^2)} = \frac{\gamma}{\beta} \times \frac{k_s^2}{k} \tag{6.40}$$

It is indicated that the online implementation complexity of the BSPCA method is less than PCA if $\tau < 1$. As β and γ are parameters which represent the dimension reduction performance of the PCA method, they should have similar values. The crucial part of Eq. (6.40) is the selected numbers of principal components. Generally, the value of parameter τ is around 1, thus the online computation complexities of BSPCA and PCA are comparative. However, the online implementation of the traditional KPCA method is much more difficult, the complexity of which can be given as

$$TO_{KPCA} = 2mkn + n + 2k \tag{6.41}$$

Similarly, if we use the relationship $\beta = m/k$, Eq. (6.41) becomes

$$TO_{KPCA} = O(n\beta k^2) \tag{6.42}$$

It is clear that the online complexity of KPCA is much higher than that of PCA and BSPCA, the relationships between these three methods can be summarized as follows

$$TO_{KPCA} \gg TO_{BSPCA} \approx TO_{PCA} \tag{6.43}$$

Therefore, the offline and online algorithm complexity analyses of the method both indicate that BSPCA is much more computationally efficient than the traditional KPCA method in monitoring nonlinear systems. Precisely, the computational complexity of BSPCA is comparative to the conventional linear PCA method, especially when it is used for online implementation. In summary, we can conclude that BSPCA can be used for nonlinear process monitoring, without increasing evident computation complexity of the algorithm.

Chapter 7
Time-Varying Process Monitoring

7.1 Introduction

Multivariate statistical process control (MSPC) schemes such as principal component analysis (PCA) and partial least squares (PLS) are widely used. There were representative and recent research efforts on continuous, batch, dynamic, multi-scale, and plant-wide processes (Kresta et al. 1991; Nomikos and MacGregor 1994, 1995a, b; Ku et al. 1995; Bakshi 1998; Thornhill and Horch 2007; AlGhazzawi and Lennox 2008; Narasimhan and Shah 2008). However, traditional MSPC schemes are formulated based on the assumption that the process variables are sampling independent, Gaussian distributed, and linearly correlated. Actually, many industrial processes are always nonlinear and their operation conditions are time-varying, due to product changes and set-point changes, among others. Furthermore, some process variables may not follow Gaussian distribution. Therefore, it is necessary to develop new methods to address these problems.

To overcome the shortcomings of MSPC schemes for nonlinear processes, several nonlinear extensions of PCA have been reported. Kramer (1992) developed a nonlinear PCA method based on the autoassociative neural network. Dong and McAvoy (1996) proposed a nonlinear PCA by combining the principal curve and the neural network. Hiden et al. (1999) used genetic programming to address the same problem. However, neural network suffers from drawbacks such as complex training strategy. Recently, a nonlinear process monitoring method based on kernel PCA (KPCA) was proposed (Lee et al. 2004; Choi et al. 2005; Cho et al. 2005). The main advantage of the KPCA method over other nonlinear PCA approaches is that no nonlinear optimization needs to be involved. Kruger et al. (2005) proposed a new nonlinearity measure for principal component and discussed the criteria on the selection of linear or nonlinear PCA for a specific process. Alternatively, Cheng and Chiu (2005) used just-in-time-learning (JITL) model with finite impulse response (FIR) structure to address the same problem. Although the adaptive FIR model can capture the dynamic of a nonlinear process, process nonlinearity might not be adequately modeled with a small number of training samples. Support vector regression (SVR), which only needs a small number of training samples, has been widely used for nonlinear

Z. Ge, Z. Song, *Multivariate Statistical Process Control,* Advances in Industrial Control, DOI 10.1007/978-1-4471-4513-4_7, © Springer-Verlag London 2013

function regression and system identification since the last decade (Vapnik 1995). Recently, Least squares support vector regression (LSSVR) was developed (Suykens et al. 2002). Because of its improved computational efficiency compared to SVM, the LSSVR model can be updated much faster, which enables online process modeling.

Motivated by the fact that MSPC is not efficient for time-varying processes, adaptive PCA and PLS methods were developed (Dayal and MacGregor 1997; Qin 1998; Li et al. 2000; Wang et al. 2003; Wang et al. 2005). Li et al. (2000) presented a monitoring strategy which built a recursive PCA (RPCA) model to update the monitoring model. To avoid fault accommodation, Wang et al. (2005) proposed an N-step-ahead horizon strategy for fault monitoring. As an alternative, model library-based methods have also been introduced in which predefined models match their corresponding modes (Zhao et al. 2004; Hwang and Han 1999; Chen and Liu 2000). However, the transition between two operating modes was always falsely alarmed. To handle this issue, several research papers have been published, including Bhagwat et al. (2003a, b) and Srinivasan et al. (2004). However, they are either model-based or too complicated to implement in practice. An overview of the transition process monitoring was given by Kourti (2005). Recently, process knowledge was incorporated into MSPC for time-varying process monitoring (Jin et al. 2006; Lee et al. 2006). Unfortunately, process knowledge is difficult to obtain from modern complex processes.

In this chapter, a local model approach for monitoring time-varying processes is introduced (Ge and Song 2008). The problem can be successfully solved through online updating of the local model. In this method, a history database which contains normal process data is needed. The LSSVR model is used to extract the nonlinear information from the raw process data. Hence, the resulting residual between the actual process output and the predicted output from the local model is no longer sensitive to process nonlinearity. In the next step, the process output and the predicted one are used in the MSPC analysis to draw monitoring charts.

Independent component analysis (ICA) is an emerging technique for finding several independent variables as linear combinations of measured variables. It can reveal more meaningful information in non-Gaussian data than PCA. Several applications of ICA have been reported (Hyvarinen and Oja 2000). Lee et al. (2004a, b) used ICA for process monitoring. Kano et al. (2004) developed a unified framework for MSPC, which combined PCA-based SPC and ICA-based SPC. A two-step information extraction strategy based on ICA-PCA was proposed by Ge and Song (2007).

The method introduced in this chapter focuses on developing an adaptive monitoring scheme for nonlinear and multiple mode processes with non-Gaussian information (Ge and Song 2008). Precisely, this chapter focuses on the situation that the changes are driven by process inputs. First, the LSSVR model is built for online modeling of the nonlinear process. Second, a local modeling strategy is utilized to handle the time-varying behavior of the process. Finally, a two-step information extraction and monitoring strategy based on ICA-PCA is used for online monitoring.

7.2 Local Modeling Strategy Based on JITL and LSSVR

7.2.1 LSSVR

SVR is a machine-learning tool and especially useful for the classification and prediction with small sample cases. This approach is motivated by statistical learning theory which leads to a class of algorithms characterized by the use of nonlinear kernels, high generalization ability, and the sparseness of the solution. Unlike the classical neural network approach, the SVR formulation of the learning problem leads to quadratic programming (QP) with linear constraint. However, the size of the matrix involved in the QP problem is directly proportional to the number of training samples. To reduce the complexity of optimization processes, LSSVR is proposed to obtain a linear set of equations instead of a QP problem in the dual space. The formulation of LSSVR is introduced as follows.

Consider a given train dataset $\{\mathbf{u}_i, y_i\}_{i=1,2,\cdots N}$, with input data $\mathbf{u}_i \in \mathbf{R}^n$ and output data $y_i \in \mathbf{R}$. The following regression model can be constructed by using nonlinear mapping function $\varphi(\cdot)$.

$$y(\mathbf{u}) = \mathbf{w}^T \varphi(\mathbf{u}) + b \tag{7.1}$$

where $\mathbf{w}$ is the weight vector and b is the bias term. By mapping the original input data into a high-dimensional space, the nonlinear separable problem becomes linearly separable in space. Then, the following cost function is formulated in the framework of empirical risk minimization

$$\min J(\mathbf{w}, b, e) = \frac{1}{2}||\mathbf{w}||_2^2 + \frac{C}{2}\sum_{\mathrm{i=1}}^{\mathrm{N}} e_i^2 \tag{7.2}$$

subject to equality constraints

$$y_i = \langle \mathbf{w}, \Phi(\mathbf{u}_i) \rangle + b + e_i, \quad i = 1, \ldots, \mathrm{N} \tag{7.3}$$

where e_i is the random error and C is a regularization parameter in determining the trade-off between minimizing the training errors and minimizing the model complexity. To solve this optimization problem, a Lagrange function is constructed as

$$L(\mathbf{w}, b, \mathbf{e}, \alpha) = J(\mathbf{w}, \mathbf{e}) - \sum_{i=1}^{N} \alpha_i \{\mathbf{w}^T \varphi(\mathbf{u}) + b + e_i - y_i\} \tag{7.4}$$

where α_i is Lagrange multiplier. The solution of Eq. (7.4) can be obtained by partially differentiating with respect to $\mathbf{w}$, b, e_i, and α_i, it also can be written as

$$\begin{bmatrix} \mathbf{0} & \mathbf{1}_{\mathrm{N}}^{\mathrm{T}} \\ \mathbf{1}_{\mathrm{N}} & \mathbf{\Omega} + \frac{1}{\mathbf{C}}\mathbf{I} \end{bmatrix} \begin{bmatrix} b \\ \alpha \end{bmatrix} = \begin{bmatrix} \mathbf{0} \\ y \end{bmatrix} \tag{7.5}$$

where

$$\mathbf{1}_\mathrm{N} = [1 \ldots 1]^\mathrm{T} \in \mathbf{R}^\mathrm{N},\ \alpha = [\alpha_1 \ldots \alpha_\mathrm{N}]^\mathrm{T} \in \mathbf{R}^\mathrm{N}, \mathbf{y} = [y_1 \ldots y_\mathrm{N}]^\mathrm{T} \in \mathbf{R}^\mathrm{N},$$
$$\mathbf{I} \in \mathbf{R}^{\mathrm{N}\times\mathrm{N}},\ \Omega_{ij} = K(\mathbf{u}_i, \mathbf{u}_j) = \langle\varphi(\mathbf{u}_i), \varphi(\mathbf{u}_j)\rangle,\ \forall i, j = 1, \ldots, \mathrm{N} \tag{7.6}$$

Finally, b and α_i can be obtained by the solution to the linear system

$$\hat{b} = \frac{\mathbf{1}^\mathbf{T}(\Omega + C^{-1}\mathbf{I}_\mathbf{N})^{-1}\mathbf{y}}{\mathbf{1}^\mathbf{T}(\Omega + C^{-1}\mathbf{I}_\mathbf{N})^{-1}\mathbf{1}} \tag{7.7}$$

$$\hat{\alpha} = (\mathbf{\Omega} + C^{-1}\mathbf{I}_\mathbf{N})^{-1}(\mathbf{y} - \mathbf{1}\hat{b}) \tag{7.8}$$

According to Mercer' s theorem, the resulting LSSVR model can be expressed as

$$f(\mathbf{u}) = \langle\mathbf{w}, \varphi(\mathbf{u})\rangle + b = \sum_{i=1}^{\mathrm{N}} \alpha_i K(\mathbf{u}_i, \mathbf{u}) + b \tag{7.9}$$

where $K(\mathrm{u}_i, \mathrm{u})$ is the nonlinear kernel function. In comparison with some other feasible kernel functions, the RBF function is a more compact supported kernel and able to reduce computational complexity of the training process and improve generalization performance of LSSVR. As a result, RBF kernel is selected as the nonlinear kernel function here. To achieve a high level of performance with LSSVR models, some parameters have to be tuned, including the regularization parameter C and the kernel parameter.

7.2.2 *Local Modeling Approach*

Conventional data-based modeling methods focused on global approaches, such as neural networks, fuzzy set methods, and other kinds of parametric models. However, when dealing with large datasets from industrial processes, traditional approaches become difficult in terms of efficient model structure determination and optimization problem formulation. Furthermore, such models are difficult to be updated online when the process operating condition changes. On the other hand, the idea of local modeling is an approach that represents a nonlinear system with a set of simple local models valid in certain operating regions. There are several well-known examples of the local modeling approach, such as neural-fuzzy network and the T-S fuzzy model. However, most local modeling approaches suffer from the drawback of requiring *a priori* knowledge to determine the partition of operating space. When this information is not available, a complicated training strategy is needed to determine both optimal model structure and parameters of the local model.

To alleviate the aforementioned problem, JITL strategy was recently developed (Cheng and Chiu 2004, 2005; Bontempi et al. 2001) for modeling nonlinear processes. The approach is based on the ideas of local modeling and database technology.

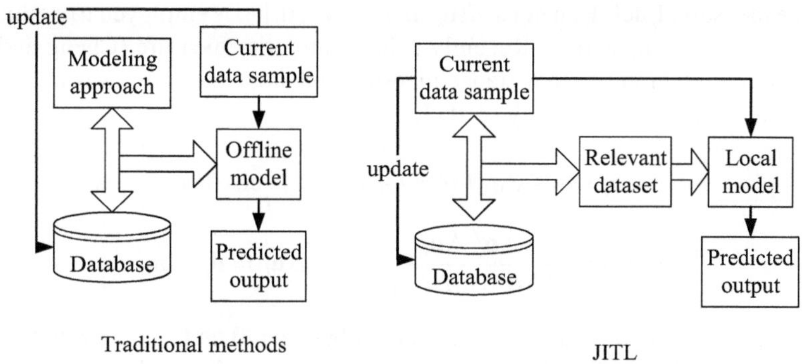

Fig. 7.1 Comparison between traditional modeling methods and JITL

JITL assumes that all available normal observations are stored in the database, and local models are built dynamically upon query. Compared to traditional modeling methods, JITL exhibits three main characteristics. First, the model development is based on the current data sample. Second, the predicted output of the current data sample is computed by the local model. Finally, the constructed local model is discarded after the predicted output is obtained. Figure 7.1 illustrates the difference between traditional global methods and JITL (Cheng and Chiu 2005; Ge and Song 2008). Traditional methods typically train the model offline, while JITL can be carried out online. When the process changes frequently, the traditional method is not only time-consuming but also will interrupt the plant operation. On the other hand, JITL shows more flexibility. Therefore, it can be used for online modeling.

There are three main steps in JITL to predict the model output corresponding to the current data sample (Cheng and Chiu 2005): (1) relevant data samples that match the current data sample are searched in the database by using some nearest neighborhood criteria; (2) a local model is built based on the relevant dataset; (3) the model output is calculated based on the local model and the current data sample. The local model is then discarded after the predicted output of the current data sample is obtained. When the next data sample comes, a new local model will be built based on the aforementioned procedures. In the previous work on JITL modeling (Cheng and Chiu 2005), an ARX/FIR model is commonly used as the local model structure. However, LSSVR is more appropriate because of the online modeling demand and its low computational cost. LSSVR is a novel machine-learning tool and is especially useful for classification and prediction with small sample cases. Details of the algorithm can be found in Suykens et al. (2002).

Having discussed the local model structure, we proceed to describe the JITL algorithm. Suppose a dataset $\{\mathbf{u}_i, \mathbf{y}_i\}_{i=1,2,\cdots N}$ is collected as the database of JITL, with input data $\mathbf{u}_i \in \mathbf{R}^n$ and output data $\mathbf{y}_i \in \mathbf{R}^m$. For a current sample data $\mathbf{u}_c$, the objective of JITL is to predict the model output $\hat{\mathbf{y}}_c = f(\mathbf{u}_c)$ according to the known database $\{\mathbf{u}_i, \mathbf{y}_i\}_{i=1,2,\cdots N}$. For the development of the local LSSVR model, the relevant data should first be selected from the database. In the previous work, the distance measure, Euclidean norm $d(\mathbf{u}_c, \mathbf{u}_i) = ||\mathbf{u}_c, \mathbf{u}_i||_2$, is employed to evaluate the similarity between $\mathbf{u}_c$ and $\mathbf{u}_i$. Recently, a new similarity measure (Cheng and Chiu 2004), which integrates both distance measure and angle measure, was proposed to improve the prediction accuracy of JITL.

$$s_i = \lambda \cdot \sqrt{e^{-d^2(\mathbf{u}_c,\mathbf{u}_i)}} + (1 - \lambda) \cdot \cos(\theta_i) \tag{7.10}$$

$$\cos(\theta_i) = \frac{\Delta\mathbf{u}_c^T \Delta\mathbf{u}_i}{||\Delta\mathbf{u}_c||_2 ||\Delta\mathbf{u}_i||_2}, \qquad \Delta\mathbf{u}_i = \mathbf{u}_i - \mathbf{u}_{i-1} \tag{7.11}$$

where λ is a weighted parameter constrained between 0 and 1, and θ_i is the angle between $\Delta\mathbf{u}_c$ and $\Delta\mathbf{u}_i$. The data will be discarded and not be used in the local model prediction if the corresponding $\cos(\theta_i)$ is negative. After all s_i are computed, they are rearranged in the descending order. In the local model construction, l relevant data $\{\mathbf{u}_i, \mathbf{y}_i\}_{i=1,2,\cdots l}$ corresponding to the l largest similarity measurements are selected, which are denoted as $(\mathbf{U}_l, \mathbf{Y}_l)$. In order to distinguish the importance of each selected sample, a weighted diagonal matrix $\mathbf{H} \in \mathbf{R}^{l \times l}$ is introduced, with diagonal elements being the l largest s_i. Therefore, input and output data are recalculated as

$$\mathbf{V}_l = \mathbf{H}\mathbf{U}_l \tag{7.12}$$

$$\mathbf{Z}_l = \mathbf{H}\mathbf{Y}_l \tag{7.13}$$

The recalculated data $(\mathbf{V}_l, \mathbf{Z}_l)$ are used as the training dataset of LSSVR. The m local LSSVR models will be built for the prediction of m outputs. The choice of the optimal value, l_{opt}, is made by minimizing the validation error

$$E_k = ||\mathbf{z}_k - f(\mathbf{v}_k)||_2 = \left|\left|\mathbf{z}_k - \sum_{i=1}^{l} \alpha_i K(\mathbf{v}_i, \mathbf{v}_k) + b\right|\right|_2 \tag{7.14}$$

where $K(\mathbf{u}_i, \mathbf{u}_j) = \exp(-||\mathbf{u}_i - \mathbf{u}_j||^2)/\sigma^2$ is chosen as a Gaussian kernel function, σ is the kernel width parameter, k is the size of the validation dataset, and α_i the Lagrange coefficient. In fact, the choice of l_{opt} should also be done under the consideration of computational burden and online modeling demand. Therefore, the selection of l_{opt} is a trade-off problem. Fortunately, the model construction of LSSVR needs less training samples than conventional methods do, and its calculation is also efficient. Consequently, the optimal predicted output for the current sample data $\mathrm{u_c}$ is calculated as (Suykens et al. 2002)

$$f(\mathbf{u}_c) = \langle \mathbf{w}, \varphi(\mathbf{u}_c) \rangle + b = \sum_{i=1}^{l_{opt}} \alpha_i K(\mathbf{u}_i, \mathbf{u}_c) + b \tag{7.15}$$

7.3 Real-Time Problem Considerations

As introduced in the last section, the local model is constructed by using JITL and LSSVR. When a new observation is available, the resembling dataset is first searched from the database. Then, a new local LSSVR is built upon the current resembling dataset. This local model-based approach is an online modeling method. Different from conventional global modeling methods, the local method is robust to the change of operation condition of the process. However, when the history database is large, searching the resembling dataset for the current observation is difficult and time-consuming. Meanwhile, the local LSSVR model should be built every time a new observation is coming, and the built local model is discarded after output predictions are obtained. In order to improve this method, the real-time problem needs to be considered (Ge and Song 2008).

Here, the newest local model, including the model structure, parameters, and the similarity limitation, which is the cutoff value of the resembling dataset used for model construction, is saved for reutilization. Thus, when a new observation is obtained, the similarity between the new observation and the most recent one is first calculated using the following similarity factor

$$s_{new} = \lambda \cdot \sqrt{e^{-d^2(\mathbf{u}_{new}, \mathbf{u}_{new-1})}} + (1 - \lambda) \cdot \cos(\theta_{new}) \tag{7.16}$$

$$\cos(\theta_{new}) = \frac{\mathbf{u}_{new}^T \mathbf{u}_{new-1}}{||\mathbf{u}_{new}||_2 ||\mathbf{u}_{new-1}||_2} \tag{7.17}$$

When the new similarity factor is not smaller than the saved similarity limitation, i.e.,

$$s_{new} \geq s_{lim} \tag{7.18}$$

The new observation is considered in the same operation mode as its most recent observation. Therefore, the saved local model can be used for output prediction of the new observation

$$f(\mathrm{u}_{new}) = \langle \mathrm{w}_r, \varphi(\mathrm{u}_{new}) \rangle + b_r = \sum_{i=1}^{l_{opt}} \alpha_i K(\mathrm{u}_i, \mathrm{u}_{new}) + b_r \tag{7.19}$$

All parameters used in Eq. (7.19) are saved. However, when the new similarity factor is smaller than the similarity limitation, i.e.,

$$s_{new} < s_{lim} \tag{7.20}$$

A new local model should be built, and the predicted output of the new observation is calculated by using Eq. (7.15). Then, the new local model is saved and the old one is updated.

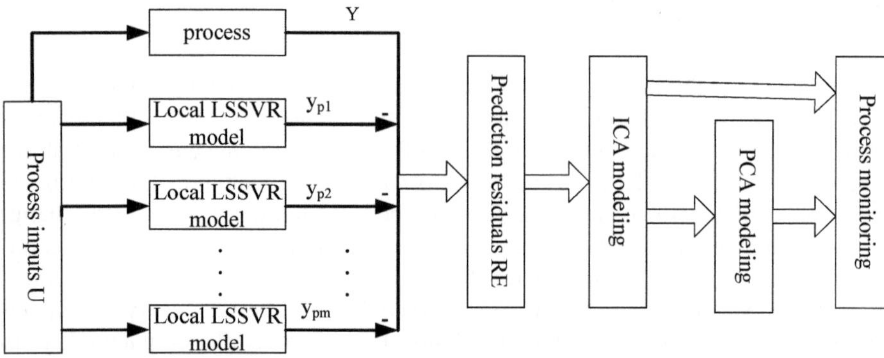

Fig. 7.2 Local LSSVR model and ICA-PCA-based monitoring scheme

7.4 Time-Varying Process Monitoring Based on Local LSSVR Model and ICA-PCA

ICA was originally proposed to solve the blind source separation problem. A two-step information extraction strategy was proposed by Ge and Song (2007), which can extract both non-Gaussian and Gaussian information. Given the dataset **X**, ICA is carried out in the first step to extract the non-Gaussian information. Then, PCA is used to model the Gaussian information. Combining the two steps, the original dataset **X** can be recalculated as (Ge and Song 2007)

$$\mathbf{X} = \mathbf{A}\hat{\mathbf{S}} + \mathbf{T}\mathbf{P}^T + \mathbf{F} \tag{7.21}$$

Having discussed the local modeling method and the information analysis strategy, we proceed to discuss the whole monitoring scheme. As shown in Fig. 7.2, each local LSSVR model is used to predict the corresponding process output. Residuals between actual process outputs and predicted ones are calculated and subsequently analyzed by ICA-PCA. Residuals can be calculated as

$$\mathbf{RE} = \mathbf{Y} - \mathbf{Y}_p \tag{7.22}$$

where $\mathbf{Y} \in \mathbf{R}^m$ is the real output, and $\mathbf{Y}_p \in \mathbf{R}^m$ is the predicted output. The next step is to use the ICA-PCA model to analyze the residual matrix, **RE**.

First, ICA is carried out. Suppose that r independent components are extracted, $\mathbf{S} = [s_1, s_2, \cdots, s_r] \in \mathbf{R}^r$. In order to monitor the non-Gaussian part of the process, the I^2 statistic is introduced as follows (Lee et al. 2004)

$$I^2 = \mathbf{s}^T\mathbf{s} \tag{7.23}$$

After the non-Gaussian information has been extracted, the residual matrix **E** is obtained. Then, PCA is used to analyze it by expanding **E** as below

$$\mathbf{E} = \sum_{i=1}^{k} \mathbf{t}_i\mathbf{p}_i^T + \mathbf{F} \tag{7.24}$$

where **F** is the residual matrix resulting from the PCA model and k is the number of selected principal components. Confidence limits of T^2 and SPE statistics to monitor the remaining Gaussian part of the process are defined below (Chiang et al. 2001)

$$T^2 = \sum_{i=1}^{k} \frac{\mathbf{t}_i^2}{\lambda_i} \leq T_{\lim}^2 = \frac{k(n-1)}{n-k} F_{k,(n-k),\alpha} \tag{7.25}$$

$$SPE = \mathbf{f}\mathbf{f}^T = \mathbf{e}(\mathbf{I} - \mathbf{P}\mathbf{P}^T)\mathbf{e}^T \leq SPE_{\lim} \tag{7.26}$$

$$\begin{aligned} SPE_{\lim} &= g\chi_{h,\alpha}^2 \\ g &= v/2m \\ h &= 2m^2/v \end{aligned} \tag{7.27}$$

where k is the number of PCs, α is significance level, c_α is the normal deviate corresponding to the upper $1-\alpha$ percentile, m and v are the mean and variance values of SPE of the training dataset, $T_{\lim}^2$ and $SPE_{\lim}$ correspond to the confidence limits of T^2 and SPE statistics, respectively.

In the PCA monitoring approach, the confidence limit is based on a specific distribution. However, in ICA monitoring, the independent component does not conform to a specific distribution. Hence, the confidence limit of the I^2 statistic cannot be determined directly from a particular approximate distribution. For simplicity, it can be determined that the number of samples outside the control limit is 1 % of the entire training samples while the process is operated under a normal condition.

Construction procedures of the local monitoring scheme are summarized as follows

1. Process data under multiple mode conditions are collected to generate the database.
2. Another process dataset under normal condition is collected. JITL algorithm is carried out and local LSSVR models are built for the prediction of the process output, and the residual is obtained.
3. ICA-PCA information extraction and analysis strategy is carried out upon the residual matrix, and the monitoring ICA-PCA model is built.
4. When a new observation comes, the similarity between the new observation and the last one is first calculated. There are two possible cases:
   (a) If the similarity exceeds the similarity limitation, the local model should be rebuilt. Thus, the resembling data are obtained and local LSSVR models are rebuilt. Also, the residual vector $\mathbf{re}_{new}$ between the actual process output and the local LSSVR model-predicted output is calculated.
   (b) If the similarity is inside the similarity limitation, the local LSSVR model does not need to be rebuilt and the process output can be obtained through the most recent local model which was saved. Thus, the residuals vector $\mathbf{re}_{new}$ can be easily obtained.

5. Scale the residual vector $\mathbf{re}_{new}$, and three monitoring statistics I^2_{new}, T^2_{new}, and SPE_{new} SPE are calculated. If these statistics are below their corresponding control limits, the process is considered to be normal. Otherwise, some process faults may have occurred.

The online process monitoring scheme of the local LSSVR model and ICA-PCA-based method is shown in Fig. 7.3.

7.5 Simulation Study

Consider a nonlinear system given by Dong and McAvoy (1996)

$$\begin{aligned} y_1 &= u + \varepsilon_1 \\ y_2 &= u^2 - 3u + \varepsilon_2 \\ y_3 &= -u^3 + 3u^2 + \varepsilon_3 \end{aligned} \tag{7.28}$$

where u is the system input, y_1, y_2, y_3 are system outputs, $\varepsilon_1, \varepsilon_2, \varepsilon_3$ are independent random noises following the distribution of $N(0, 0.01)$. The system input is $u \in [0.01,\ 1]$ for one operating mode and $u \in [0.7,\ 1.2]$ for another. Two hundred process data are generated for each of the operating mode. To simulate time-varying and mode transition behaviors of the process, a ramp signal with an increment of 0.005 was added to u after the first operating mode is carried out. These data are stored as the database of the system. Then, another 200 process data under normal operating mode are generated to construct local LSSVR models and the ICA-PCA model is built upon the prediction residual. Parameters chosen in modeling procedures are: $l = 25, C = 25, \lambda = 0.5$, the number of independent components (IC) is selected as 1, and the number of principal components (PC) is also 1. In order to compare the result with conventional methods, the linear JITL model ($l = 100$), the traditional RPCA model and the KPCA model are also built. Monitoring results are analyzed as follows.

Figure 7.4 shows monitoring results of the normal process, without operating mode changes. The dashed line represents 99 % confident control limit. Good monitoring results are given by all four methods. It can be inferred that the new method does not lose the monitoring efficiency for processes in normal operation. When the process is time-varying and experiencing operating mode changes, the poor performance of the KPCA model can be clearly observed in Fig. 7.5d. On the other hand, as demonstrated in Fig. 7.5a, the local method can monitor the time-varying and multiple operating mode process effectively with a small number of samples lying in the vicinity of the 99 % control limit. Monitoring results of the other two methods are given in Fig. 7.5b and 7.5c, respectively. In comparison to the nonlinear local method, the linear JITL method results in poor monitoring performance because during the time-varying process, there are not enough training samples (similar to the current sample) for linear JITL modeling. Although the number of training samples is set as 100, most of them are not similar to the current sample. On the other

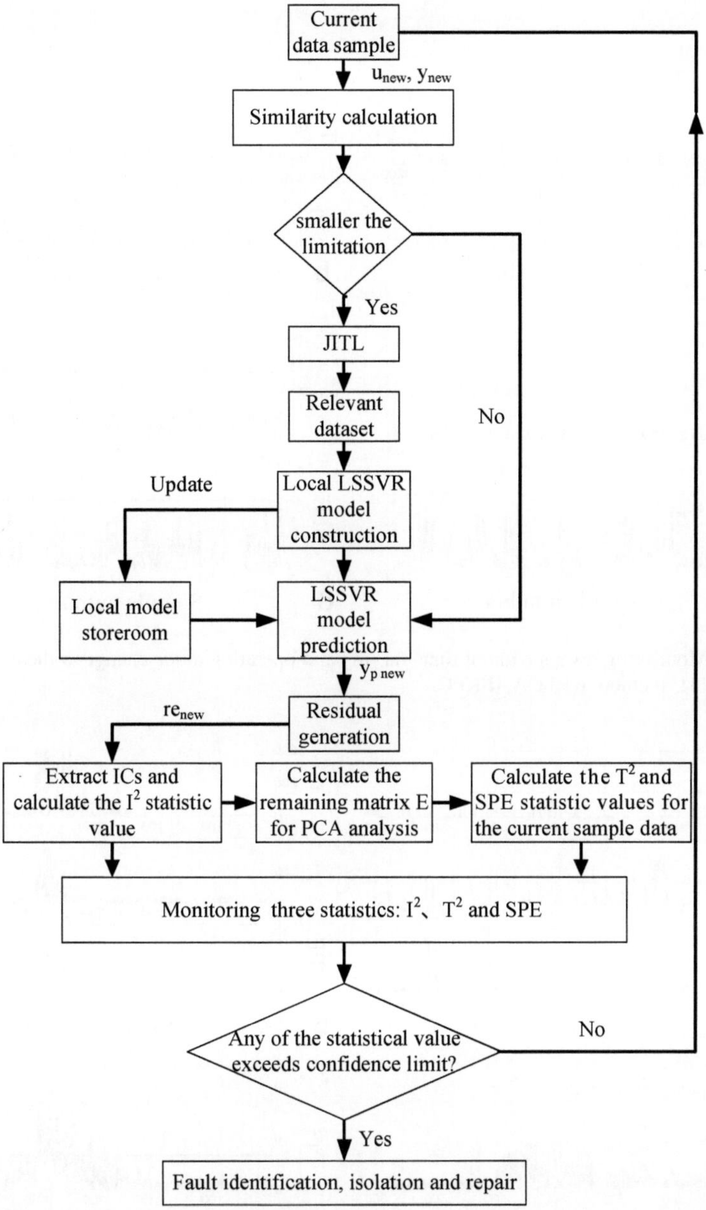

Fig. 7.3 Online process monitoring scheme based on local LSSVR model and ICA-PCA

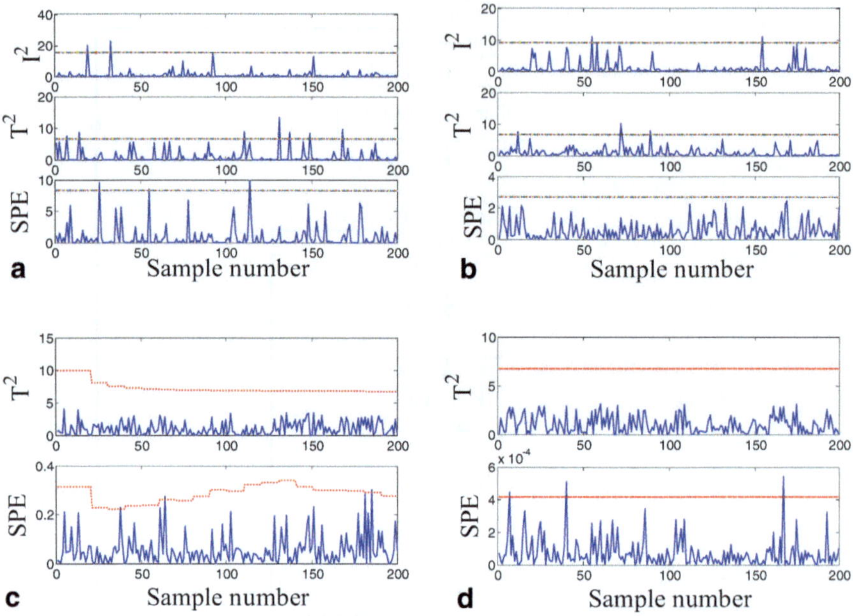

Fig. 7.4 Monitoring results without time-varying and operating mode change. **a** the new method, **b** linear JITL method, **c** RPCA, **d** KPCA

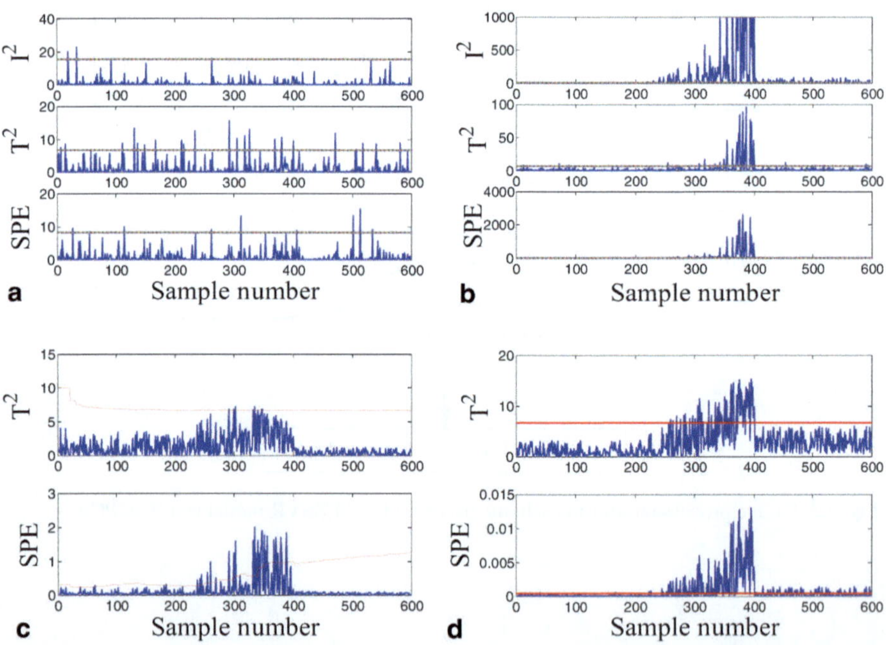

Fig. 7.5 Monitoring results with time-varying and operating mode change. **a** the new method, **b** linear JITL method, **c** RPCA, **d** KPCA

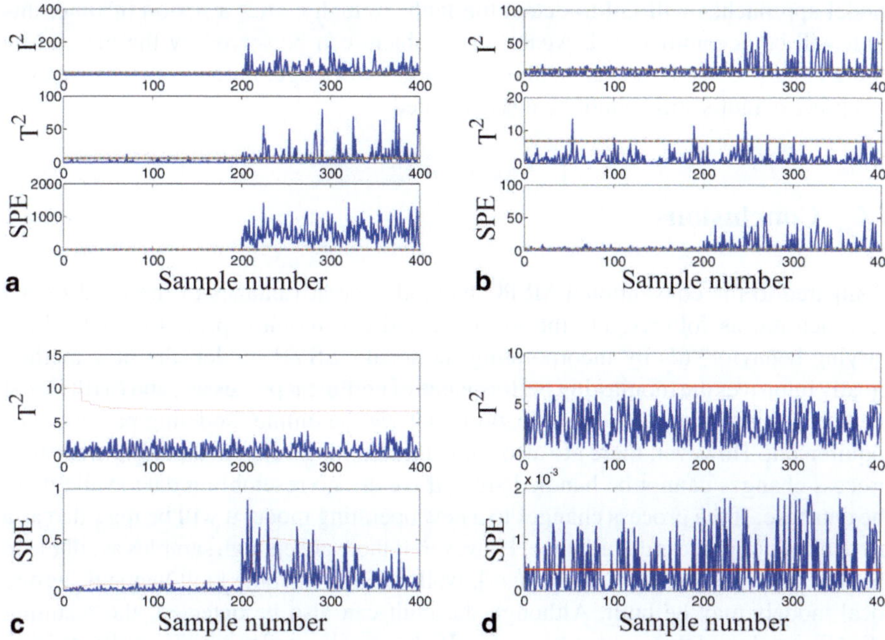

Fig. 7.6 Monitoring results of the fault. **a** the new method, **b** linear JITL method, **c** RPCA, **d** KPCA

hand, the number of training samples needed for the new method is much smaller, which results in the building of a model with improved accuracy. The RPCA model is adaptive during process mode transition. However, there are still several false alarms triggered during this period of the process.

Next, a fault is introduced in the process. The fault is a small deviation occurred in y_2 and is described by

$$y_1 = u + \varepsilon_1$$
$$y_2 = 1.1u^2 - 3u + \varepsilon_2$$
$$y_3 = -u^3 + 3u^2 + \varepsilon_3 \quad (7.29)$$

Monitoring results of this fault are shown in Fig. 7.6. Among the four methods, three of them can successfully detect the fault, except the RPCA method. As RPCA is an adaptive method, it cannot identify whether the process change is normal or abnormal. To address this issue, one can refer to Wang et al. (2005). Because the database of the new method or linear JITL method is assumed to contain normal process data only, the resembling dataset to the current faulty sample for the local model (LSSVR or linear) building cannot represent the current process behavior, which will result in wrong predicted output of the current sample. Therefore, the residual between the actual output and the predicted output is considered to be abnormal. However, if the abnormal samples are also collected and saved in the same database, both local model approaches will fail to detect the fault. Actually, after a period of time, this fault will be accommodated. Again, the problem can be solved by the method by Wang et al. (2005). However, without process knowledge, normal process changes and process faults still cannot be discriminated.

7.6 Conclusions

Compared to the conventional MSPC methods, the advantages of the local model approach are as follows: (1) the local method can monitor processes with time-varying behavior; (2) by incorporating the local LSSVR model, the new method greatly improves the monitoring performance of nonlinear processes; and (3) the local method is capable to, with modifications, satisfy the online modeling performance requirement. However, there are also some limitations of this method. First, the new process changes cannot be handled since there are no resembling data available in the database. If the process changes to a new operating mode, it will be regarded as a certain fault or process disturbance. However, if there are enough samples available in the database, the method will still work well. Second, if some fault happens, wrong local models may be built. Although the fault can also be detected, the resulting LSSVR local models may not be useful. If abnormal samples are also collected and saved in the database, the fault will be accommodated.

Moreover, there are several open questions that should be considered in future works: (1) selection of the size of the resembling dataset. Although a small number of training samples are available for the LSSVR model construction, the size of the resembling dataset should be carefully determined to achieve better modeling performance and (2) choice of the weighted parameter λ, which is selected empirically here. Besides, identification and discrimination of operating condition changes and process faults are also important issues in the process monitoring area. Without process knowledge, these tasks are always very difficult. However, the development of data-mining technologies and knowledge engineering provides useful insight for future work on this topic.

Chapter 8
Multimode Process Monitoring: Part 1

8.1 Introduction

As a critical approach to the safe and efficient operation of a wide range of industrial processes, multivariate statistical process control (MSPC) has been intensively researched in the last several decades. Particularly, principal component analysis (PCA) and partial least squares (PLS) have been widely used for process monitoring and fault diagnosis (Kruger et al. 2001; Wang et al. 2004; Simoglou et al. 2005; Cherry and Qin 2006; AlGhazzawi and Lennox 2008; Kruger and Dimitriadis 2008). More recently, independent component analysis (ICA) was employed for monitoring processes with non-Gaussian information, which can effectively extract high-order statistic information from process data (Li and Wang 2002; Lee et al. 2004, 2006; Albazzaz and Wang 2007; Zhao et al. 2008b). For those processes which are mainly driven by non-Gaussian latent variables, ICA was considered to be more appropriate for information extraction and thus enhanced the sequent monitoring performance. However, rather than being solely driven by Gaussian or non-Gaussian latent variables, some processes may simultaneously be driven by these two latent variables. To this end, some combinations of PCA and ICA methods have been developed. The first combination of ICA and PCA was proposed by Kano et al. (2004a). Recently, a different two-step ICA-PCA-based combination method was proposed by Ge and Song (2007) and later adopted by Zhao et al. (2008a) in monitoring batch processes, which can combine the advantages of both methods for feature extraction purpose.

However, for process monitoring, it is difficult to apply the traditional MSPC methods in multimode processes, since the current techniques are based on the assumption that the process has only one nominal operating region. Therefore, problems will arise when those techniques are applied to processes with multiple operating modes. In fact, industrial processes may change frequently due to fluctuations in raw materials, set point changes, etc. The application of traditional MSPC techniques to a process with multiple operating modes may cause continuous false alarms, even when the process is under another steady-state nominal operating mode. To solve this problem, recursive or adaptive PCA and PLS methods have been proposed (Qin 1998; Li et al. 2000; Wang et al. 2005). However, these approaches are carried out blindly, which means that continuous updating is performed whether a process change has

Z. Ge, Z. Song, *Multivariate Statistical Process Control*, Advances in Industrial Control,
DOI 10.1007/978-1-4471-4513-4_8,

been identified (or not). Therefore, false alarms may be caused during the transition between two operation modes. Alternatively, model library-based methods have been introduced for multimode process monitoring (Hwang and Han 1999; Zhao et al. 2004; Singhai and Seborg 2006). Predefined models match their corresponding operating modes. Similarly, the transition between two operating modes was always false alarmed. To partition the process data softly, the Gaussian mixture model has also been introduced (Chen and Liu 1999, 2000; Choi et al. 2004; Yu and Qin 2008; Chen and Sun 2009). However, the data distribution inside each operation mode was still assumed to be Gaussian. More recently, process knowledge is incorporated into MSPC for monitoring time-varying and multimode processes, including the external analysis method, adaptive local modeling, etc (Jin et al. 2006; Ge and Song 2008a; Kano et al. 2004b; Ge et al. 2008b). Unfortunately, process knowledge is difficult to obtain in many processes. To this end, a Bayesian-based process monitoring method has been developed (Ge and Song 2009). First, the two-step ICA-PCA algorithm is used for dimension reduction and information extraction. Clustering methods are then employed for unsupervised monitoring of multimode processes. In order to construct the method in a probabilistic manner and monitor different operation modes softly, traditional monitoring statistics are transferred to fault probabilities, followed by the Bayesian inference.

After some fault has been detected, a fault identification can be developed under the same monitoring framework (Ge and Song 2009). Various methods for fault identification have been developed in recent years (Westerhuis et al. 2000; Dunia and Qin 1998; Lieftuche et al. 2006; Krzanowski 1979; Johannesmeyer et al. 2002; Raich and Cinar 1997). To our knowledge, most of them can be treated as a pattern classification problem. Thus, all previous faulty datasets were first collected, and then modeled and stored as faulty scenarios. When a fault has been detected, it will be matched to all the known faults in the storeroom, where the most resemble one is voted to represent the current detected fault for further analysis. In this monitoring method, when a fault has been detected, the posterior probability of this fault to each fault scenario in the storeroom will be calculated, depending on which fault identification is carried out.

8.2 Mode Clustering and Feature Extraction

In order to develop the monitoring method through a probabilistic manner, each data sample should be assigned to different operation modes, with the corresponding responsibility to each mode. Then, the fault probability of the data sample in each modeled operation zone can be calculated. Without any useful information or expert knowledge of the process, unsupervised learning method can be adopted. Here, we assume that the operation modes have not been identified yet, thus all process data are mixed together. To implement the algorithm unsupervised, data clustering methods such as K-means, Fuzzy c-means, etc. can be employed (Sebzalli and Wang 2001; Chen and Wang 2009; Bicego and Figueiredo 2009). In this section, the mode clustering and feature extraction algorithms are introduced, which is followed by the Bayesian monitoring and fault identification method in the next section.

8.2.1 Fuzzy C-Mean Method for Operation Mode Clustering

Based on the theory of fuzzy sets, the Fuzzy c-means clustering method has been developed and applied to many areas. The ability of this fuzzy set theory is to express in numerical format the impression that stems from a large group of elements into clusters that do not have sharply defined boundaries. The membership of an object in the fuzzy set is given by a subjectively defined fuzzy membership function, whose value can range from zero to one. Suppose the process consists of C operation modes, each of which contains n_c $(c = 1, 2, \dots, C)$ objects. Therefore, the total data samples can be summarized as $n = \sum_{c=1}^{C} n_c$. Given these n data patterns, $\mathbf{x}_1, \mathbf{x}_2, \dots, \mathbf{x}_n$ each of which is described by m features, thus $\mathbf{x}_i \in R^m (i = 1, 2, \dots, n)$. The aim of Fuzzy c-means is to partition the data patterns into C clusters, which correspond to the C operation modes of the process. This clustering process is defined by finding cluster centers $\mathbf{z}_1, \mathbf{z}_2, \dots, \mathbf{z}_c$, which minimizes the following fuzzy performance index (Sebzalli and Wang 2001):

$$f(\mathbf{Z}) = \sum_{i=1}^{n} \sum_{c=1}^{C} \Gamma_{ic} |\mathbf{x}_i - \mathbf{z}_c| \tag{8.1}$$

where $\mathbf{Z}$ consists of C vectors of cluster centers. Γ_{ic} is defined as

$$\Gamma_{ic} = \left[\sum_{j=1}^{C} \left(\frac{d_{ic}}{d_{ij}} \right)^{2/(\beta-1)} \right]^{-1}, \quad 1 \le i \le n, \quad 1 \le c \le C,$$

$$d_{ij} = ||\mathbf{x}_i - \mathbf{z}_c||, \quad 1 \le i \le n, \quad 1 \le j \le C \tag{8.2}$$

Generally, the Fuzzy c-means algorithm is an interactive procedure. Hence, after several iterative steps, the original process dataset can be partitioned into subgroups, which we referred to operation modes in the process.

8.2.2 Two-Step ICA-PCA Feature Extraction for Each Operation Mode

As most processes consist of a large amount of variables, dimension reduction techniques are often necessary for feature extraction. Among them, PCA is one of the most widely used techniques. However, it only depends on the second-order statistics, high-order statistical information cannot be well extracted by this method. Comparatively, ICA was more efficient for extracting high-order data information, while simultaneously keeping the extracted components as independent as possible. It has been examined that the combination of ICA and PCA performs better than either of these two methods, especially when they were applied for process monitoring. While ICA/PCA can only monitor the non-Gaussian/Gaussian information of the process, the two-step ICA-PCA method can simultaneously monitor those two kinds of information, which enhances the performance. Specifically, both the non-Gaussian and Gaussian variations can be captured by the two-step ICA-PCA method.

After whole process data have been partitioned into C clusters, which we represent as $\mathbf{X} = [\mathbf{X}_1|\mathbf{X}_2| \cdots |\mathbf{X}_C]$, the two-step ICA-PCA method is carried out upon each subgroup dataset $\mathbf{X}_c$ $(c = 1, 2, \ldots, C)$. ICA is carried out in the first step to extract the high-order and non-Gaussian information, therefore $\mathbf{X}_c$ could be decomposed as follows (Ge and Song 2007):

$$\begin{aligned} \mathbf{X}_c &= \mathbf{A}_c\hat{\mathbf{S}}_c + \mathbf{X}'_c \\ \hat{\mathbf{S}}_c &= \mathbf{W}_c\mathbf{X}_c \\ \mathbf{X}'_c &= \mathbf{X}_c - \mathbf{A}_c \cdot \hat{\mathbf{S}}_c \end{aligned} \tag{8.3}$$

where $\mathbf{A}_c$ and $\mathbf{W}_c$ are the mixing and demixing matrices, $\mathbf{X}'_c$ is the residual matrix after the ICA step. In the second step, PCA is used to model the Gaussian information in the residual matrix $\mathbf{X}'_c$. Thus, $\mathbf{X}'_c$ is decomposed as

$$\mathbf{X}'_c = \mathbf{T}_c\mathbf{P}_c^T + \mathbf{R}_c \tag{8.4}$$

where $\mathbf{T}_c$ and $\mathbf{P}_c$ are score and loading matrices of the PCA decomposition and $\mathbf{R}_c$ is the residual matrix after the analysis of PCA.

Through the analysis of the two-step ICA-PCA method, a critical issue is how to determine the retained numbers of independent components (ICs) and principal components (PCs). Some selection methods for the number of PCs have already been outlined in several researches (Valle et al. 1999; Wang et al. 2002), while recently a new method for determination of the number of ICs has been illustrated (Liu et al. 2008). Hence, depending on the existing methods, the numbers of both ICA and PCA can be determined appropriately, which ensures the monitoring performance of the sequent step.

8.3 Bayesian-Based Method for Process Monitoring and Fault Identification

In this section, the Bayesian monitoring method is detailed, followed by the fault identification algorithm. Finally, some analyses and remarks are given.

8.3.1 Multimode Process Monitoring Through the Bayesian Viewpoint

Traditionally, after both the Gaussian and non-Gaussian information have been extracted, several monitoring statistics such as I^2, T^2, and SPE as well as their corresponding control limits can be established for monitoring purpose, which are given as follows (Ge and Song 2007):

$$I_c^2 = \hat{\mathbf{s}}_c^T\hat{\mathbf{s}}_c \leq I_{c,\lim}^2 \tag{8.5}$$

$$T_c^2 = \sum_{i=1}^{k_{pc}} \frac{\mathbf{t}_i^2}{\lambda_i} \leq T_{c,\lim}^2 = \frac{k_{pc}(n-1)}{n-k_{pc}} F_{k_{pc},(n-k_{pc}),\alpha} \tag{8.6}$$

$$SPE_c = \mathbf{r}_c \mathbf{r}_c^T \leq SPE_{c,\lim} = g_c \chi^2_{h_c,\alpha} \tag{8.7}$$

where $c = 1, 2, \ldots, C$, k_{pc} is the number of PCs, α is the selected significance level, and $g_c = v_c/(2m_c)$, $h_c = 2m_c^2/v_c$, in which m_c and v_c are the mean and variance values of *SPE* within operation mode *c*. The confidence limit of the I^2 statistic can be determined simply by trial-and-error or precisely by kernel density estimation (Chen et al. 2004).

After the individual monitoring model has been built for each operation mode, they can be used for online monitoring of new data samples. However, when a new data sample $\mathbf{x}_{new}$ has been obtained, we have no information which operation mode it belongs to, and which model should be used to monitor this data sample. It could be problematic when an inappropriate model is used for monitoring, because false alarms may be caused if the data sample does not belong to the corresponding operation mode. Therefore, we intend to implement the algorithm in a probabilistic framework, which can softly assign the new data sample to different operation modes. In this framework, the traditional monitoring task is transferred to a probabilistic viewpoint, which is based on the Bayesian inference (Bishop 2006). First, the values of traditional monitoring statistics I^2, T^2, and *SPE* are transferred to probability values in each operation mode, which are given as follows (Ge and Song 2009):

$$P_{I^2}(\mathbf{x}_{new}|c) = \exp\left\{-\frac{\kappa I^2_{c,new}(\mathbf{x}_{new})}{I^2_{c,\lim}}\right\} \tag{8.8}$$

$$P_{T^2}(\mathbf{x}_{new}|c) = \exp\left\{-\frac{\kappa T^2_{c,new}(\mathbf{x}_{new})}{T^2_{c,\lim}}\right\} \tag{8.9}$$

$$P_{SPE}(\mathbf{x}_{new}|c) = \exp\left\{-\frac{\kappa SPE_{c,\lim}(\mathbf{x}_{new})}{SPE_{c,\lim}}\right\} \tag{8.10}$$

where $c = 1, 2, \ldots, C, \kappa$, is a tuning parameter, $I^2_{c,new}(\mathbf{x}_{new})$, $T^2_{c,new}(\mathbf{x}_{new})$, and $SPE_{c,new}(\mathbf{x}_{new})$ can be calculated through Eqs. (8.3–8.7). Through the Bayesian inference, the posterior probabilities of each operation mode corresponding to the three monitoring statistics are given as

$$P_{I^2}(c|\mathbf{x}_{new}) = \frac{P_{I^2}(c, \mathbf{x}_{new})}{P_{I^2}(\mathbf{x}_{new})} = \frac{P_{I^2}(\mathbf{x}_{new}|c)P(c)}{\sum_{c=1}^{C}[P_{I^2}(\mathbf{x}_{new}|c)P(c)]} \tag{8.11}$$

$$P_{T^2}(c|\mathbf{x}_{new}) = \frac{P_{T^2}(c, \mathbf{x}_{new})}{P_{T^2}(\mathbf{x}_{new})} = \frac{P_{T^2}(\mathbf{x}_{new}|c)P(c)}{\sum_{c=1}^{C}[P_{T^2}(\mathbf{x}_{new}|c)P(c)]} \tag{8.12}$$

$$P_{SPE}(c|\mathbf{x}_{new}) = \frac{P_{SPE}(c, \mathbf{x}_{new})}{P_{SPE}(\mathbf{x}_{new})} = \frac{P_{SPE}(\mathbf{x}_{new}|c)P(c)}{\sum_{c=1}^{C}[P_{SPE}(\mathbf{x}_{new}|c)P(c)]} \tag{8.13}$$

where $P(c)$, $c = 1, 2, \ldots, C$ are prior probabilities, which can be simply determined as

$$P(c) = \frac{n_c}{n} \tag{8.14}$$

where n_c is the number of training samples in operation mode c, and n is the total number of data samples of the process.

When the posterior probability of the new data sample $\mathbf{x}_{new}$ has been obtained, we are in the position to determine whether this sample is normal or not. To this end, another three probabilities correlated to fault detection are defined as follows:

$$P^c_{f,I^2}(\mathbf{x}_{new}) = \Pr\{I^2_c(\mathbf{x}_{tr,c}) \le I^2_c(\mathbf{x}_{new})\} \tag{8.15}$$

$$P^c_{f,T^2}(\mathbf{x}_{new}) = \Pr\{T^2_c(\mathbf{x}_{tr,c}) \le T^2_c(\mathbf{x}_{new})\} \tag{8.16}$$

$$P^c_{f,SPE}(\mathbf{x}_{new}) = \Pr\{SPE_c(\mathbf{x}_{tr,c}) \le SPE_c(\mathbf{x}_{new})\} \tag{8.17}$$

where $c = 1, 2, \ldots, C$, $\mathbf{x}_{tr,c}$ is the training sample in operation mode c. The values of these three probabilities can be simply determined by counting the number the training samples whose statistic values are small than that of the new data sample. However, if we want to determine these values more precisely, the resampling method or ensemble method can be employed. Generally, these techniques can be referred in Bishop (2006), Ho (1998), and Breiman (1996). Therefore, provided that the new data sample belongs to each operation mode, these probabilities are indications of whether this new monitored data sample is normal or not.

Under the consideration of the new data sample belonging to multiple operation modes, a new Bayesian monitoring statistic can be constructed based on the posterior probabilities and the fault probabilities as

$$BMS_{I^2}(\mathbf{x}_{new}) = \sum_{c=1}^{C} [P_{I^2}(c|\mathbf{x}_{new}) P^c_{f,I^2}(\mathbf{x}_{new})] \tag{8.18}$$

$$BMS_{T^2}(\mathbf{x}_{new}) = \sum_{c=1}^{C} [P_{T^2}(c|\mathbf{x}_{new}) P^c_{f,T^2}(\mathbf{x}_{new})] \tag{8.19}$$

$$BMS_{SPE}(\mathbf{x}_{new}) = \sum_{c=1}^{C} [P_{SPE}(c|\mathbf{x}_{new}) P^c_{f,SPE}(\mathbf{x}_{new})] \tag{8.20}$$

In contrast to the hard assignment of the data sample into one operation mode, the soft assignment can avoid the potential risk of false alarm induced by misassignment. As the values of $P^c_{f,I^2}(\mathbf{x}_{new})$, $P^c_{f,T^2}(\mathbf{x}_{new})$, and $P^c_{f,SPE}(\mathbf{x}_{new})$ are all ranged from zero to one, the bounds of all Bayesian monitoring statistics are also between zero and one.

8.3.2 Fault Identification and New Operation Mode Determination

When a fault has been successfully detected, the fault identification issue is considered. Suppose there are totally F faulty operation modes in the process, the detected faulty data sample is denoted as $\mathbf{x}_{fault}$. Similarly, based on the Bayesian inference, the posterior probability of the faulty sample belongs to a specific faulty mode f that can be calculated as (Ge and Song 2009)

$$P_{I^2}(f|\mathbf{x}_{fault}) = \frac{P_{I^2}(f,\mathbf{x}_{fault})}{P_{I^2}(\mathbf{x}_{fault})} = \frac{P_{I^2}(\mathbf{x}_{fault}|f)P(f)}{\sum_{c=1}^{F}[P_{I^2}(\mathbf{x}_{fault}|f)P(f)]} \tag{8.21}$$

$$P_{T^2}(f|\mathbf{x}_{fault}) = \frac{P_{T^2}(f,\mathbf{x}_{fault})}{P_{T^2}(\mathbf{x}_{fault})} = \frac{P_{T^2}(\mathbf{x}_{fault}|f)P(f)}{\sum_{c=1}^{F}[P_{T^2}(\mathbf{x}_{fault}|f)P(f)]} \tag{8.22}$$

$$P_{SPE}(f|\mathbf{x}_{fault}) = \frac{P_{SPE}(f,\mathbf{x}_{fault})}{P_{SPE}(\mathbf{x}_{fault})} = \frac{P_{SPE}(\mathbf{x}_{fault}|f)P(f)}{\sum_{c=1}^{F}[P_{SPE}(\mathbf{x}_{fault}|f)P(f)]} \tag{8.23}$$

where $f = 1,2,\ldots,F$, the items in Eqs. (8.21–8.23) can be calculated through procedures developed in the previous subsection. To simplify the fault identification task, these three posterior probabilities can be combined together, with appropriate weighted parameters as

$$P(f|\mathbf{x}_{fault}) = \alpha P_{I^2}(f|\mathbf{x}_{fault}) + \beta P_{T^2}(f|\mathbf{x}_{fault}) + \gamma P_{SPE}(f|\mathbf{x}_{fault}) \tag{8.24}$$

To identify the type of the detected fault, a voting strategy based on the maxima value of the posterior probability could be used. To be straightforward, when some faults are similar in pattern storeroom, this voting strategy may fail, because two or more posterior probability values may be comparative. However, the value of posterior probability does give important information for fault identification. If a new fault has been detected, the posterior probability values of all fault scenarios will be very small. In this case, a new faulty scenario is considered to be identified. If enough data samples have been collected, the new model should be built and this new faulty pattern can be added to the pattern storeroom. However, based on the maxima voting strategy, the new fault may be assigned to its most similar fault scenario in the pattern storeroom. To avoid this pitfall, joint probabilities of the faulty sample and all known faulty scenarios can be used, which are defined as follows:

$$\begin{aligned} &P(f,\mathbf{x}_{fault}) = P(\mathbf{x}_{fault}|f)P(f) \\ &f = 1,2,\ldots,F \end{aligned} \tag{8.25}$$

All joint probabilities will approach to zero when a new fault is detected. Meanwhile, if a new operation mode is carried out in the process, it can also be identified by these joint probabilities.

8.3.3 Remarks

For those multimode processes whose data are Gaussian distributed, the Gaussian mixture model can be employed for monitoring. However, the Gaussian mixture model-based method may cause singularity problem, since the estimated variance matrix may be ill-conditioned due to colinearity between process variables. On the other side, the Bayesian-based method does not have this problem. Through the two-step ICA-PCA method, the data information can be well extracted and the dimension of process variables can also be reduced. Besides, process variables are de-correlated, which avoids the singularity problem that may happen in the Gaussian mixture model approach. Based on the Bayesian inference theory, the monitored sample can be assigned to different operation modes. By transferring traditional monitoring statistics into probability distribution values, monitoring results in different operation modes can be easily combined to form a final decision for the monitored data sample. In this monitoring framework, three monitoring statistics can be resorted, which can greatly enhance the fault detection sensitivity. Another remarkable issue is the differentiation between the normal operation mode change and the new process faulty mode. Although both these kind of new operation modes can be identified by the proposed method, they can not be easily distinguished without any process or expert knowledge.

8.4 Illustrations and Results

In this section, two case studies are given for evaluation of the Bayesian-based monitoring method.

8.4.1 A Numerical Example

This simple system is given as follows:

$$\begin{bmatrix} x_1 \\ x_2 \\ x_3 \\ x_4 \\ x_5 \end{bmatrix} = \begin{bmatrix} 0.5768 & 0.3766 \\ 0.7382 & 0.0566 \\ 0.8291 & 0.4009 \\ 0.6519 & 0.2070 \\ 0.3972 & 0.8045 \end{bmatrix} \begin{bmatrix} s_1 \\ s_2 \end{bmatrix} + \begin{bmatrix} e_1 \\ e_2 \\ e_3 \\ e_4 \\ e_5 \end{bmatrix} \tag{8.26}$$

where $\begin{bmatrix} s_1 & s_2 \end{bmatrix}^T$ are considered as latent variables or sources, and $[e_1 \ \ e_2 \ \ e_3 \ \ e_4 \ \ e_5 \ ]^{\mathrm{T}}$ are zero-mean white noises with standard deviation of 0.01. Multiple operating modes are constructed by different data sources, which are given as follows:

$$\text{First mode}: \begin{array}{l} s_1(k): \ \mathit{Uniform}(-10,-7); \\ s_2(k): \ N(-5,1) \end{array}$$

$$\text{Second mode}: \begin{array}{l} s_1(k): 2\cos(0.08k)\sin(0.006k); \\ s_2(k): \ \mathrm{sign}\,[\sin(0.03k)+9\cos(0.01k)] \end{array}$$

$$\text{Third mode}: \begin{array}{l} s_1(k): \ \mathit{Uniform}(2,5); \\ s_2(k): \ N(7,1) \end{array}$$

where k is a sample index. Two hundred samples of each operating mode are generated by the above equations. Therefore, a total of 600 training samples are used for process monitoring model constructions. For testing purpose, three datasets were generated with each one having 200 samples, which are listed as follows.

Case 1 The system is running at the first mode.
Case 2 The system is initially running at the first mode, and then a step change of 1 is added to variable 3 starting from sample 101 to the end of the operation.
Case 3 The system is initially running at mode 3, and then a ramp change of variable 1 from sample 101–200 by adding $0.02(i - 100)$ to each sample value in this range, where i is the sample number.

To build the monitoring model, two independent components and one principal component have been selected. After the two-step ICA-PCA feature extraction step, the Bayesian-based monitoring model can be developed by following the steps described in the last section. First, the normal process dataset is tested. The monitoring results through the Bayesian monitoring statistic (BMS) are shown in Fig. 8.1a, which indicates that no fault has been detected through this testing. For comparison, we then monitor this process dataset by the single ICA-PCA model, which was constructed under each of the operation mode. Because the testing dataset was generated under the first operation mode, the monitoring results obtained by the first single ICA-PCA model are expected to be good, which are given in Fig. 8.1b. However, due to the usage of inappropriate single models, this testing dataset is judged to be abnormal by the second and third single ICA-PCA models in Fig. 8.2c, d. Similar monitoring results can be obtained by single PCA models that were developed under multiple operation modes. The difference between the single ICA-PCA and PCA models lies in the previous feature extraction strategy. In order to identify which operation mode the current testing dataset belongs to, the posterior probability or the joint probability can be used. The joint probabilities corresponding to each operation mode are plotted in Fig. 8.2. By examining their values, the testing dataset should be identified as the first operation mode, which is the real case.

To evaluate the fault detection ability of the Bayesian method, two generated faulty datasets are then tested. Their corresponding type I and II errors are listed in Table 8.1, which are mean values of 100 process runs. The following shows some realizations

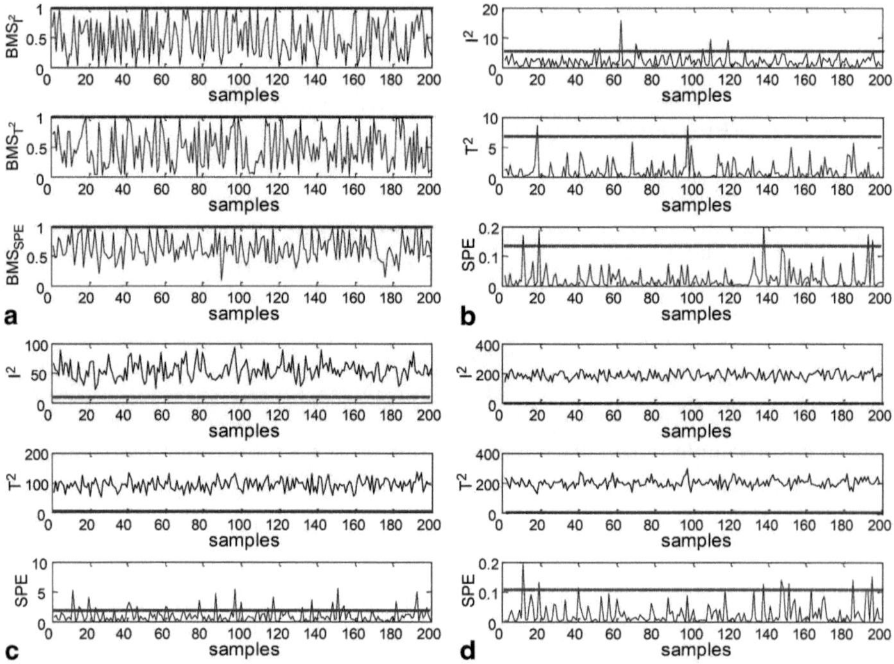

Fig. 8.1 Monitoring results of the normal process. **a** BMS monitoring statistics. **b** First single ICA-PCA model. **c** Second single ICA-PCA model. **d** Third single ICA-PCA model

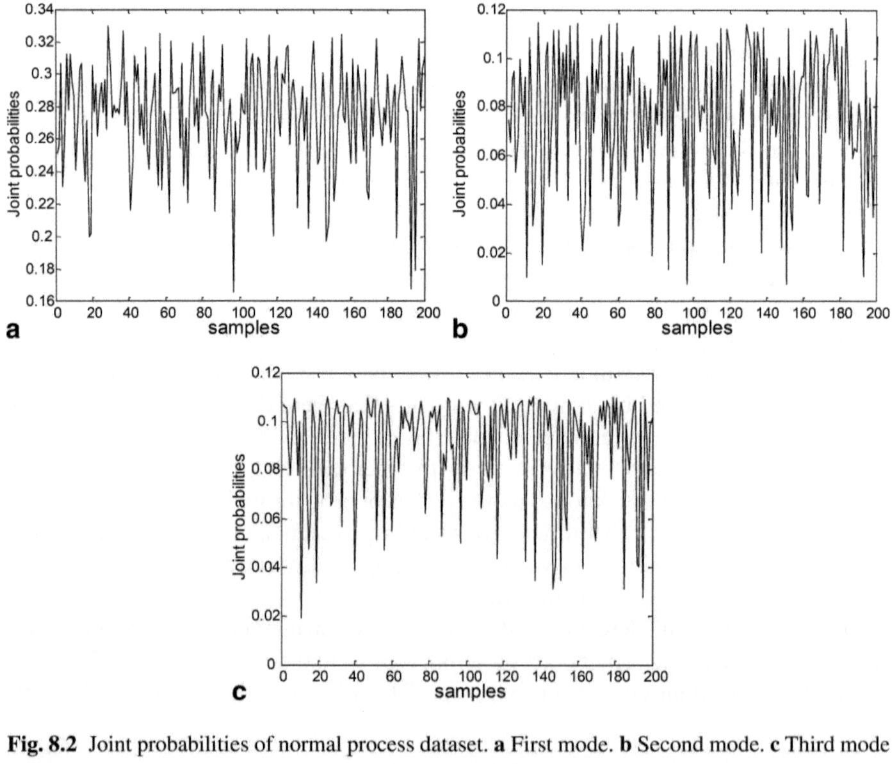

Fig. 8.2 Joint probabilities of normal process dataset. **a** First mode. **b** Second mode. **c** Third mode

Table 8.1 Type I and II errors of two faults in simple example process

Faults/errors	Type I			Type II		
	BIF_I^2	BIF_T^2	BIF_SPE	BIF_I^2	BIF_T^2	BIF_SPE
Fault 1	0.0024	0.0020	0.0006	0	0	0
Fault 2	0.0086	0.0013	0.0006	0.0139	0.0118	0.0085

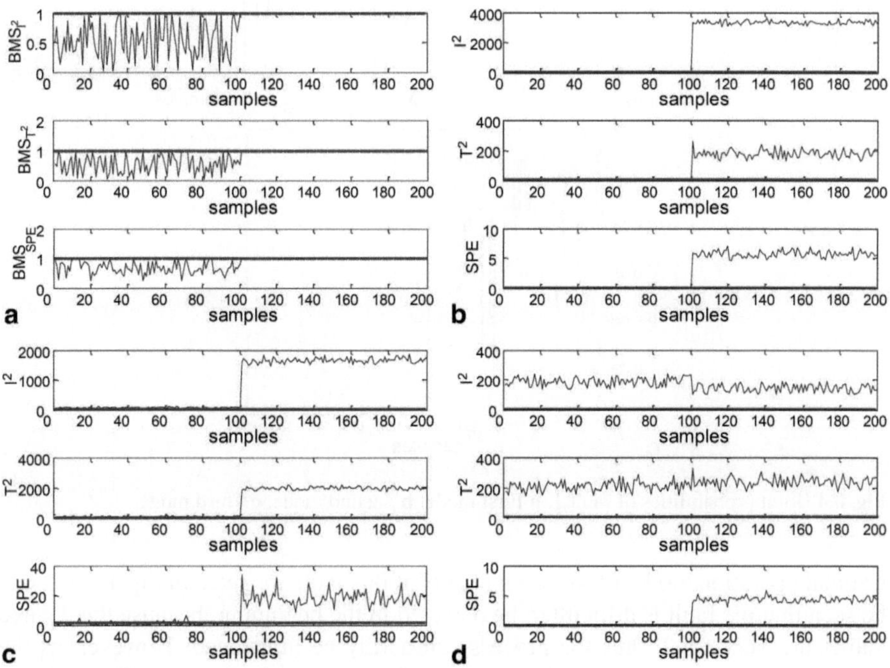

Fig. 8.3 Monitoring results of fault 1. **a** BMS monitoring statistics. **b** First single ICA-PCA model. **c** Second single ICA-PCA model. **d** Third single ICA-PCA model

of their monitoring results. As shown in Fig. 8.3, three BMS statistics successfully detected this fault suddenly after the fault was introduced, which indicates that the proposed Bayesian method is very sensitive to step changes of the process. Compared to the monitoring results of single ICA-PCA methods in Fig. 8.3b–d, one can find that the fault can be detected by the first single ICA-PCA model. Tracking the joint probability changes of this fault, the process is judged to be at the first operation mode during the first 100 samples, and then the operation mode changes, since the joint probabilities approach to zero after sample 101, which are shown in Fig. 8.4. Different from the first fault, the second one is a ramp fault. This process first ran at the third operation mode, and then a slow change was introduced to one of the process variables. This slow change can cause slow drifts of the process from the

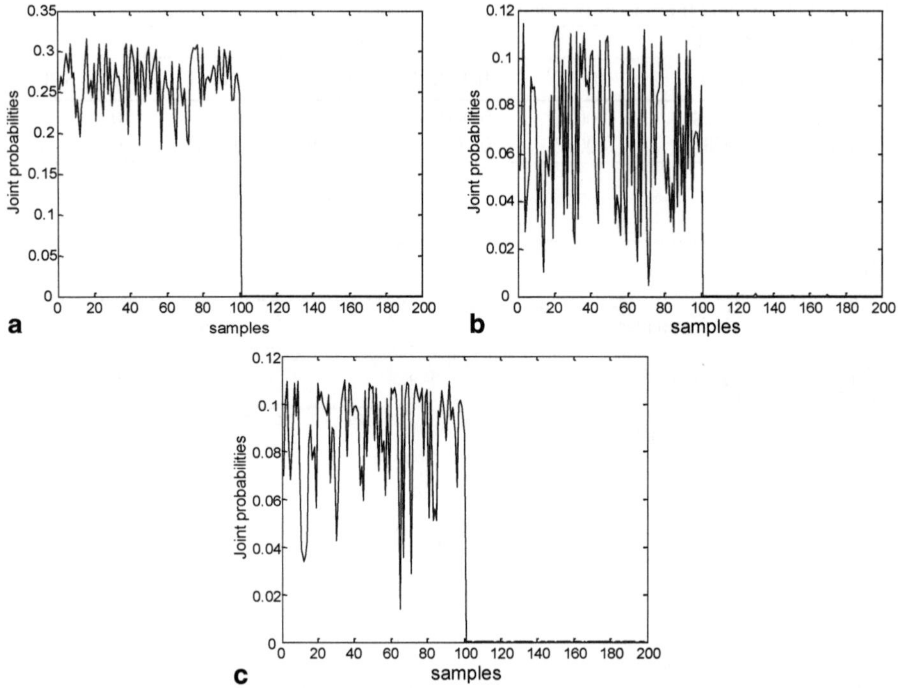

Fig. 8.4 Joint probabilities of fault 1. **a** First mode. **b** Second mode. **c** Third mode

original operation mode. Monitoring results of this fault are given in Fig. 8.5a. It can be seen that the fault is difficult to be detected in the beginning, because the change values are very small thus the process can hardly be influenced. However, as the fault becomes more and more obvious, it can be successfully detected. Similarly, by examining the joint probability analyses results in Fig. 8.5b–d, it can be inferred that this faulty dataset was first generated under the third operation mode, and then the fault was introduced.

8.4.2 TE Benchmark Case Study

The details of the TE process description can be found in Chap. 3 and Downs and Vogel (1993). Again, 16 continuous process variables are selected for process monitoring. There are six operation modes available for simulation case studies, two of which (mode 1 and mode 3) are selected for the simulation of multimode processes. In the present study, we denote these two modes as operation mode 1 and operation mode 2. It is noted that through the simulation study, the process was shut down when fault 6 was introduced. Therefore, fault 6 was excluded from this study.

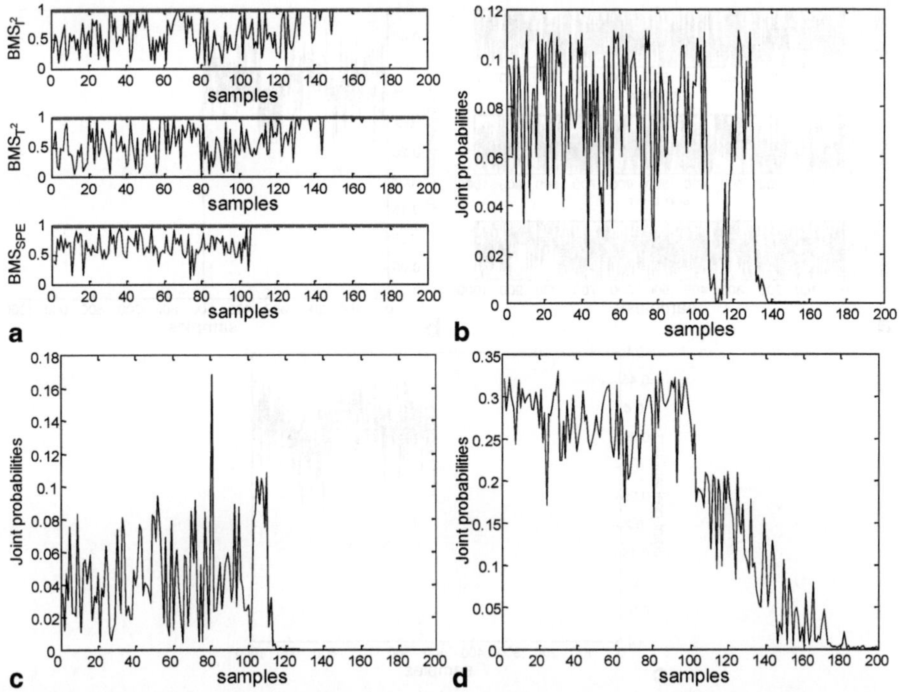

Fig. 8.5 Monitoring and joint probability analysis results of fault 2. **a** BMS monitoring statistics. **b** Joint probabilities in the first mode. **c** Joint probabilities in the second mode. **d** Joint probabilities in the third mode

The simulation datasets which we have collected were separated into two parts: the training datasets and the testing datasets, both of which consisted of 1,000 observations. Their sampling interval was 3.6 min. All process faults were introduced after sample 500, thus they were either run upon operation mode 1 or carried out upon mode 2. To develop the Bayesian monitoring model, three independent components and six principal components are chosen. All confidence limits are selected as 99 %. To evaluate the monitoring performance of the proposed method for multimode processes, two datasets of mode 1 and mode 3 were joined together, thus 2,000 samples represented for normal operation conditions are first tested. The monitoring results of the Bayesian monitoring statistics are given in Fig. 8.6a, which gives normal information for these two normal datasets. To identify which operation mode the current monitoring sample belongs to, the joint probability analysis can again be employed, results of which are shown in Fig. 8.6b, c. It can be inferred that the first 1,000 samples belong to operation mode 1, and the last 1,000 samples belong to operation mode 2, which is in accordance with the real characteristic of the testing datasets.

Next, some selected process faults are tested. Fault 1 is a step change of stream 4 in the process. The process initially runs under operation mode 1, and then the

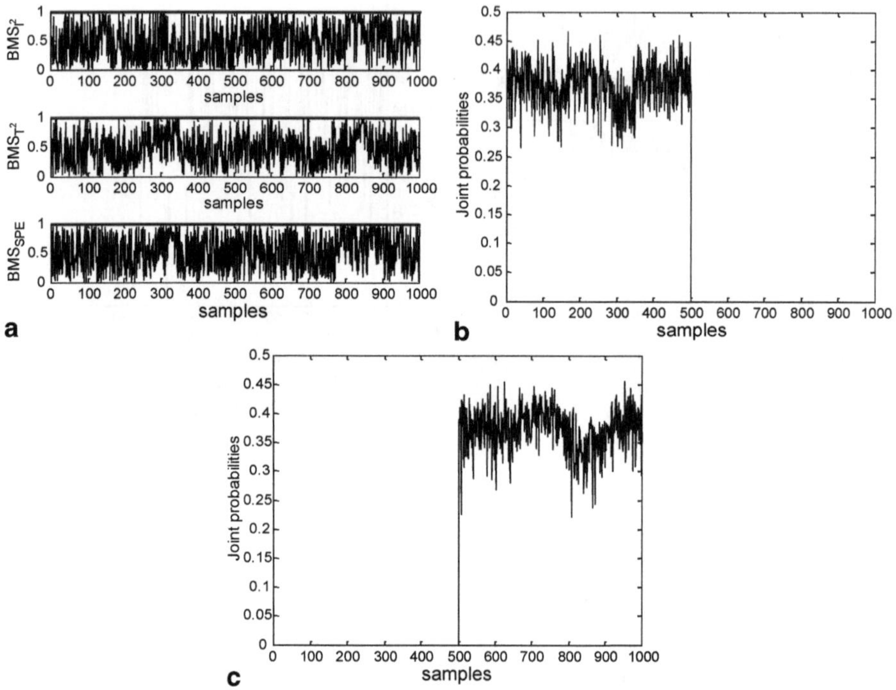

Fig. 8.6 Monitoring and Joint probability analysis results of normal operation modes. **a** BMS monitoring statistics. **b** Joint probabilities in the first mode. **c** Joint probabilities in the second mode

Table 8.2 Type I and II errors of two process faults in TE process

Faults/errors	Type I			Type II		
	BIF_I^2	BIF_T^2	BIF_SPE	BIF_I^2	BIF_T^2	BIF_SPE
Fault 1	0.0147	0.0401	0.0162	0.0144	0.0076	0.0038
Fault 10	0.0150	0.0397	0.0152	0.0844	0.0532	0.0258

fault is introduced at sample 501, a total of 1,000 samples are generated. Similarly, 1,000 testing samples are generated when the process initially runs under operation mode 2. Joining these two datasets together, a new 2,000 samples dataset is obtained. Then, 100 simulations are carried out, the mean values of type I and II errors of these process faults are tabulated in Table 8.2. As one of these realizations, the successful monitoring results are given in Fig. 8.7a. All Bayesian monitoring statistics exceed their corresponding limits when the fault was introduced. When the process runs under normal operation conditions, the values of these Bayesian monitoring statistics return back under their confidence limits. Joint probability plots of this process are given in Fig. 8.7b,c, which indicates that the process initially ran under operation

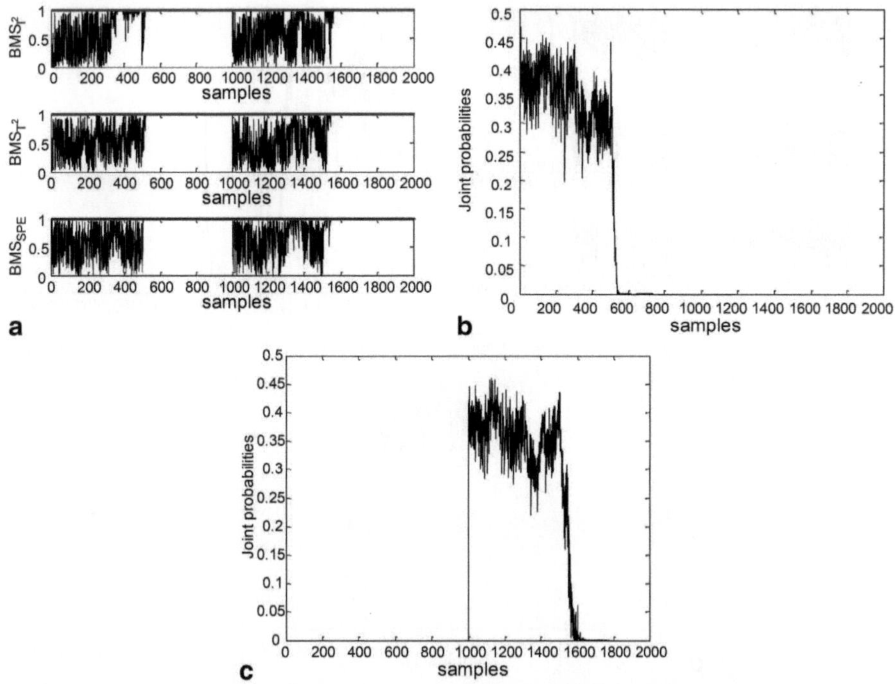

Fig. 8.7 Monitoring and joint probability analysis results of fault 1. **a** BMS monitoring statistics. **b** Joint probabilities in the first mode. **c** Joint probabilities in the second mode

mode 1, and then the process changed since the joint probability values approach to zero from sample 501 to sample 1,000. Similarly, fault 10 is tested, which is a random temperature change of the C feed in the TE process. A same number of test samples are generated through the same manner of fault 1. Monitoring and joint probability analysis results are given in Fig. 8.8. It is clear that two process changes can be found, which correspond to the fault introduced under the two operation modes starting from sample 501.

When the fault has been successfully detected, we can further identify it. Previously, a fault pattern storeroom should be built. Thus, a total of 19 fault patterns have been constructed and stored for fault identification. One thousand faulty samples were generated under each faulty operation condition. To identify fault 1 that has been successfully detected, the posterior probability can be used to deem which fault pattern it belongs to. The posterior probability values of all 1,000 samples of this fault correspond to fault pattern 1 are plotted in Fig. 8.9a. It can be found that most values are close to one, which indicated that the correct fault pattern has been identified. In comparison, the posterior probability values correspond to other 18 fault patterns are all close to zero. Here, only the results of fault 10 are shown in Fig. 8.9b. Similarly, the fault identification results of fault 10 are given in Fig. 8.10.

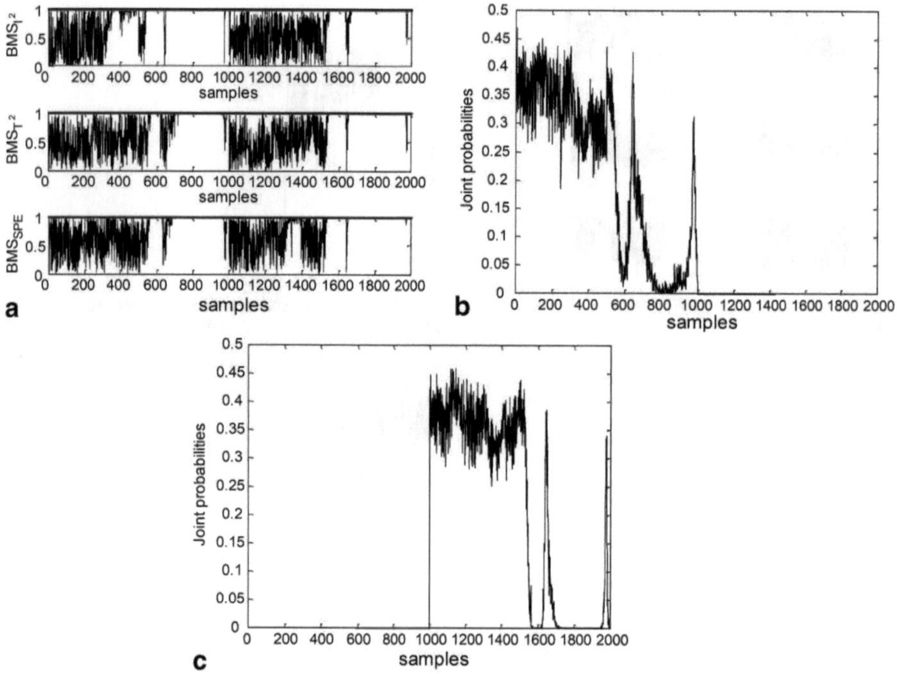

Fig. 8.8 Monitoring and joint probability analysis results of fault 10. **a** BMS monitoring statistics. **b** Joint probabilities in the first mode. **c** Joint probabilities in the second mode

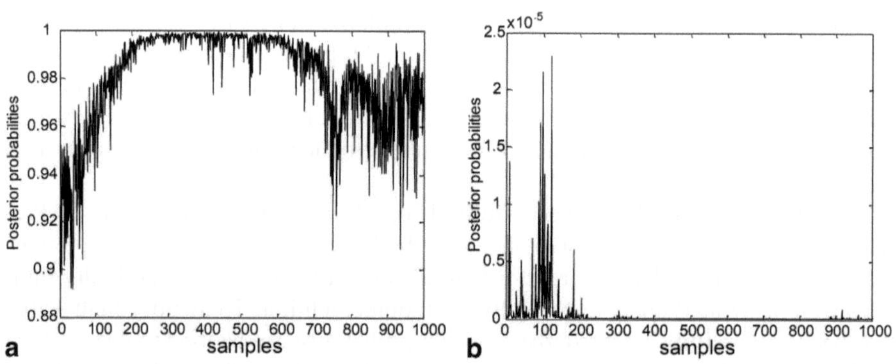

Fig. 8.9 Posterior probabilities of fault 1 corresponding to **a** Fault 1 and **b** Fault 10

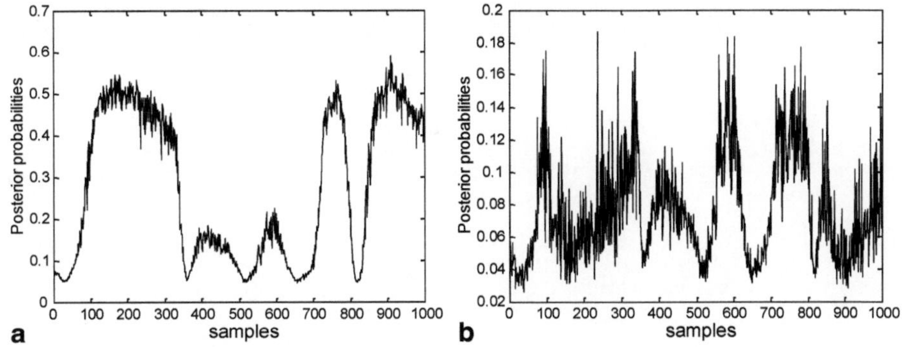

Fig. 8.10 Posterior probabilities of fault 10 corresponding to **a** fault 1 and **b** fault 10

8.5 Conclusions

In the present chapter, a Bayesian-based method has been introduced for monitoring multimode processes. To reduce the dimension of the process data, the two-step ICA-PCA feature extraction method was used, which can facilitate the sequent monitoring and analysis steps. By transferring traditional monitoring statistic values to fault probabilities in each operation mode, the monitoring results in different operation modes can be easily combined by the Bayesian inference strategy. Compared to the mixed Gaussian models, the singularity problem was avoided in this method. Through analyses of the posterior probability and the joint probability for the monitored data sample, more information has been extracted, thus the process monitoring could be more reliable. Like other fault identification approaches, the new fault identification scheme is also a pattern classification method, which takes the most resembling fault scenario as the identification decision for the detected fault. To test the monitoring and fault identification performance of the proposed method, three case studies have been carried out, which all showed its feasibility and efficiency.

Fig. 5.10 Posterior probabilities of [illegible]

Chapter 9
Multimode Process Monitoring: Part 2

9.1 Introduction

With growing requirements of safety and product quality in modern industrial processes, monitoring and fault diagnosis have become increasingly important. Data-based process methods such as multivariate statistical process control (MSPC) have gained much attention. MSPC-based monitoring methods such as principal component analysis (PCA) and partial least squares (PLS) have been widely reported both in academic and industrial areas (Chiang et al. 2001; Qin 2003; Singhal and Seborg 2006). Traditional MSPC methods are under the assumption that the process is linear, single mode, Gaussian, and stationary. However, most industrial processes are nonlinear, multimode, non-Gaussian, and dynamic. Many extensions have been developed to the traditional MSPC monitoring methods for processes with particular characteristics (Chen and Chen 2006; Lee et al. 2004; Thissen et al. 2005; AlGhazzawi and Lennox 2008; Hu and Yuan 2008; Nomikos and MacGregor 1995b; Yao and Gao 2008).

This chapter is focused on the nonlinear and multimode characteristics of the process. While the nonlinear characteristic inherently exists in the real process, the diversity requirement of the modern market necessitates the multimode behavior of the process. Compared to the linear case, faults are always more difficult to be detected in nonlinear processes, because the correlations between process variables are much more complex. So far, to our knowledge, only several nonlinear process monitoring methods have been developed, such as principal curve, neural network, kernel PCA, etc. (Dong and McAvoy 1996; Hiden et al. 1999; Choi et al. 2005; Maulud et al. 2006; Wang et al. 2007; Zhang and Qin 2008; Zhao et al. 2009). The nonlinear transformations of these developed methods make them to be computationally inefficient and difficult to use for online modeling and monitoring. On the other hand, many multimode process monitoring methods have also been developed in the last few years, including adaptive method, multiple model method, Gaussian mixture model-based method, external analysis, etc. (Qin 1998; Hwang and Han 1999; Wang et al. 2005; Choi et al. 2004; Yu and Qin 2008; Chen and Sun 2009; Ge and Song 2008; Ge et al. 2008). The implementation of the adaptive method can

Z. Ge, Z. Song, *Multivariate Statistical Process Control,* Advances in Industrial Control,
DOI 10.1007/978-1-4471-4513-4_9, © Springer-Verlag London 2013

cause false alarms, especially during the transition of two operation modes. Similar problem may arise in the multiple model-based methods, as they always assign the monitored data sample to a single monitoring model. Although the Gaussian mixed model-based method can avoid this problem, they are only developed for linear processes. Without appropriate process knowledge, the external analysis-based method is relatively inefficient for monitoring. While separate research studies have already been carried out for nonlinear or multimode processes, monitoring approaches for processes with both these behaviors have rarely been reported.

Recently, an efficient monitoring method for processes with both nonlinear and multimode characteristics has been developed (Ge and Song 2011), which can be carried out automatically and is also easy to implement online. First, the recorded process dataset is partitioned through the sample direction, which renders to multiple subgroup dataset representing different operation modes. Second, within each operation mode, instead of constructing a complicated nonlinear monitoring model, a series of linear models are developed to approximate the nonlinearity of the process. A key issue of this method is how to divide the process variable into linear subspaces. In this chapter, a two-step variable selection method is introduced for data partition through the variable direction. Third, to combine the monitoring results in different operation modes and linear subspaces, a two-dimensional Bayesian monitoring approach is formulated. Compared to traditional monitoring methods, this method cannot only reduce the computational complex of the algorithm, but also it can improve the monitoring performance for the process. With the introduced Bayesian inference and combination strategy, the decision making for nonlinear multimode processes becomes much easier.

9.2 Mode Partition and Linear Subspace Construction

In this section, process data are firstly partitioned into subgroups, which correspond to different operation modes. Then, each subgroup dataset is further partitioned into several different linear subspaces, to approximately represent the nonlinearity of the process. The structure of the process data is given in Fig. 9.1. Denote the original dataset as $\mathbf{X} \in R^{n \times m}$, where n is the total sample number, and m is the number of process variables.

9.2.1 Data Partition Through Sample Direction

As shown in Fig. 9.1, through the sample direction, the process data can be divided into several different subgroups, representing different operation modes. If some process knowledge of mode information is available, the classification of the process operation mode is straightforward. With limited process knowledge, the partition

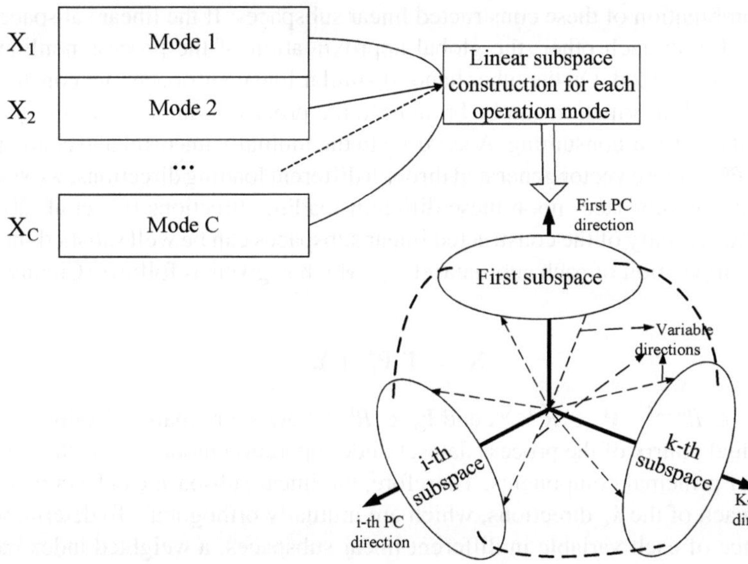

Fig. 9.1 The structure of the process dataset

of the process data may also be carried out automatically. To this end, many unsupervised clustering methods can be used, such as K-means, Fuzzy-C means, their variants, etc. (Sebzalli and Wang 2001; Chen and Wang 2009; Bicego and Figueiredo 2009). Besides, we can also use some data visualization methods, which are always implemented upon PCA or Fisher discriminative analysis (FDA). Through analysis of the data representation in the first two or three dimensions of the featured space, we can determine the mode behavior of the original dataset. For simplicity, the K-means method can be used. Suppose the original process dataset X was collected under C different operation modes, the partitioned subgroup datasets can be represented as

$$\mathbf{X} = [\mathbf{X}_1^T, \mathbf{X}_2^T, \ldots, \mathbf{X}_c^T, \ldots, \mathbf{X}_C^T]^T \tag{9.1}$$

where $c = 1, 2, \ldots, C$, $\mathbf{X}_c^T \in R^{n_c \times m}$, and $\sum_{i=1}^{C} \boldsymbol{n}_i = n$.

9.2.2 Two-Step Variable Selection for Linear Subspace Construction

After the dataset has been divided into different operation modes, we are in the position to construct linear subspaces. In this chapter, a two-step variable selection strategy is proposed for linear subspace construction. In our opinion, the diversity of the linear subspace should be intensified, as the process nonlinearity is approximated by the combination of these constructed linear subspaces. If the linear subspaces are overlapped with each other, the global approximation of the process nonlinearity cannot be guaranteed. On the other hand, if similar linear subspaces are constructed, less additional information can be brought to the process, which makes the method inefficient and time-consuming. According to the mutually uncorrelated characteristic of the PCA score vector generated through different loading directions, we suggest to build linear subspaces upon these different loading directions (Ge et al. 2010b). Hence, the diversity of the constructed linear subspaces can be well satisfied through PCA decomposition of each sub-dataset X_c, which is given as follows (Chiang et al. 2001)

$$\mathbf{X}_c = \mathbf{T}_c\mathbf{P}_c^T + \mathbf{E}_c \tag{9.2}$$

where $\mathbf{T}_c \in R^{n_c \times k_c}$, $\mathbf{P}_c \in R^{m \times k_c}$, and $\mathbf{E}_c \in R^{n_c \times m}$ are score matrix, loading matrix, and residual matrix of the process dataset under operation mode c, k_c is the selected number of principal components. Therefore, the linear subspace can be constructed through each of the k_c directions, which are mutually orthogonal. To determine the importance of each variable in different linear subspaces, a weighted index can be defined as follows

$$WI_{ij}^c = |\mathbf{P}_c(i, j)| \tag{9.3}$$

where $i = 1, 2, \ldots, m$, $j = 1, 2, \ldots, k_c$, $\mathbf{P}_c(i, j)$ is the corresponding element of the loading matrix P_c. Therefore, in the first-step selection, appropriate number of variables with large weighted index can be selected in each linear subspace, thus the new sub-dataset can be represented as $\mathbf{X}_{c,s} \in R^{n_c \times m_s}$, where m_s is the selected variable number in the first step, $c = 1, 2, \ldots, C$, $s = 1, 2, \ldots, k_c$.

In the second step, linearity of these selected variables is evaluated, depending on which are further screened. Here, correlation analysis is introduced to measure the correlation degrees among these variables, which is given as

$$\rho(i, j) = \frac{\mathrm{cov}(\mathbf{X}_{c,s}(i), \mathbf{X}_{c,s}(j))}{\sqrt{D(\mathbf{X}_{c,s}(i))} \cdot \sqrt{D(\mathbf{X}_{c,s}(j))}} \tag{9.4}$$

where $1 \leq i, j \leq m_s$, $D(\cdot)$ is the variance of the variable. The correlation degree of one variable can be measured by the correlation coefficients between this variable and remaining variables. The summation of these correlation coefficients can be used for variable selection in the second step, which is given as (Ge et al. 2011)

$$sum\rho(i) = \sum_{j=1}^{m_s} \rho(i, j) \tag{9.5}$$

where $1 \leq i \leq m_s$. Therefore, those variables which are more linear correlated are selected for linear subspace construction. Suppose the selected variable number is m_s' in each linear subspace, the final selected dataset in each linear subspace can be represented as $\mathbf{X'}_{c,s} \in R^{n_c \times m'_s} (c = 1, 2, \ldots, C,\ s = 1, 2, \ldots, k_c)$.

9.3 Two-Dimensional Bayesian Inference for Process Monitoring

When the original process dataset has been partitioned into different operation modes and the linear subspaces have also been constructed, monitoring models can be firstly developed in these linear subspaces, and then combined through the Bayesian strategy. As the monitoring result in each linear subspace can also be obtained by the Bayesian inference, a two-dimensional Bayesian statistical monitoring approach can be developed. In this section, the new method is demonstrated and the computational complexity of the method is analyzed. Besides, some remarks are also given.

9.3.1 Method Description and Implementation

First, a PCA model is developed in each linear subspace, and two traditional monitoring statistics T^2 and SPE are constructed. Thus, the linear subspace PCA model can be given as

$$\mathbf{X}'_{c,s} = \mathbf{T}_{c,s}\mathbf{P}^T_{c,s} + \mathbf{E}_{c,s} \tag{9.6}$$

where $c = 1, 2, \ldots, C$, $s = 1, 2, \ldots, k_c$, the number of principal components $k_{c,s}$ in each linear subspace can be selected differently from each other. Then, the T^2 and SPE statistics can be constructed and their confidence limits can be determined as follows (Chiang et al. 2001)

$$T^2_{c,s,\lim} = \frac{k_{c,s}(n_c - 1)}{n_c - k_{c,s}} F_{k_{c,s},(n_c - k_{c,s}),\alpha} \tag{9.7}$$

$$SPE_{c,s,\lim} = g_{c,s}\chi^2_{h_{c,s},\alpha} \tag{9.8}$$

where $g_{c,s} = v_{c,s}/(2m_{c,s})$, $h_{c,s} = 2m^2_{c,s}/v_{c,s}$, $m_{c,s}$ and $v_{c,s}$ are the mean and variance values of SPE in each linear subspace, $k_{c,s}$ is the number of PCs in the linear subspace, α is significance level, and c_α is the normal deviate corresponding to the upper $1-\alpha$ percentile.

When monitoring for a new data sample x_{new}, two statistic values $T^2_s(\mathbf{x}_{new})$ and $SPE_s(\mathbf{x}_{new})$ are first calculated in each linear subspace. In order to combine the linear subspace monitoring results in different operation modes, two transformations can be made, which are given as

$$P^s_{T^2}(\mathbf{x}_{new}|c) = \exp\left\{-\frac{T^2_s(\mathbf{x}_{new})}{T^2_{c,s,\lim}}\right\} \tag{9.9}$$

$$P^s_{SPE}(\mathbf{x}_{new}|c) = \exp\left\{-\frac{SPE_s(\mathbf{x}_{new})}{SPE_{c,s,\lim}}\right\} \tag{9.10}$$

where $s = 1, 2, \ldots, k_c$. Through the Bayesian inference (Bishop 2006), the posterior probabilities of each operation mode corresponding to each linear subspace are given as

$$P_{T^2}^s(c|\mathbf{x}_{new}) = \frac{P_{T^2}^s(c, \mathbf{x}_{new})}{P_{T^2}^s(\mathbf{x}_{new})} = \frac{P_{T^2}^s(\mathbf{x}_{new}|c)P(c)}{\sum_{c=1}^{C} [P_{T^2}^s(\mathbf{x}_{new}|c)P(c)]} \tag{9.11}$$

$$P_{SPE}^s(c|\mathbf{x}_{new}) = \frac{P_{SPE}^s(c, \mathbf{x}_{new})}{P_{SPE}^s(\mathbf{x}_{new})} = \frac{P_{SPE}^s(\mathbf{x}_{new}|c)P(c)}{\sum_{c=1}^{C} [P_{SPE}^s(\mathbf{x}_{new}|c)P(c)]} \tag{9.12}$$

where $P(c)$, $c = 1, 2, \ldots, C$ are prior probabilities, which can be simply determined as

$$P(c) = \frac{n_c}{n} \tag{9.13}$$

After the posterior probability of the monitored data sample under each operation mode has been determined, we intend to calculate their fault probabilities. Again, the Bayesian inference strategy can be employed, through which the fault probabilities in each linear subspace under different operation modes can be calculated as (Ge et al. 2011)

$$P_{f,T^2}^{s,c}(AB|\mathbf{x}_{new}) = \frac{P_{f,T^2}^{s,c}(\mathbf{x}_{new}|AB)P_{f,T^2}^{s,c}(AB)}{P_{f,T^2}^{s,c}(\mathbf{x}_{new})} \tag{9.14}$$

$$P_{f,T^2}^{s,c}(\mathbf{x}_{new}) = P_{f,T^2}^{s,c}(\mathbf{x}_{new}|N)P_{f,T^2}^{s,c}(N) + P_{f,T^2}^{s,c}(\mathbf{x}_{new}|AB)P_{f,T^2}^{s,c}(AB)$$

$$P_{f,SPE}^{s,c}(AB|\mathbf{x}_{new}) = \frac{P_{f,SPE}^{s,c}(\mathbf{x}_{new}|AB)P_{f,SPE}^{s,c}(AB)}{P_{f,SPE}^{s,c}(\mathbf{x}_{new})} \tag{9.15}$$

$$P_{f,SPE}^{s,c}(\mathbf{x}_{new}) = P_{f,SPE}^{s,c}(\mathbf{x}_{new}|N)P_{f,SPE}^{s,c}(N) + P_{f,SPE}^{s,c}(\mathbf{x}_{new}|AB)P_{f,SPE}^{s,c}(AB)$$

where

$$P_{f,T^2}^{s,c}(\mathbf{x}_{new}|AB) = \exp\left\{-\frac{T_{c,s,\lim}^2}{T_s^2(\mathbf{x}_{new})}\right\} \tag{9.16}$$

$$P_{f,SPE}^{s,c}(\mathbf{x}_{new}|AB) = \exp\left\{-\frac{SPE_{c,s,\lim}}{SPE_s(\mathbf{x}_{new})}\right\}$$

$$P_{f,T^2}^{s,c}(\mathbf{x}_{new}|N) = \exp\left\{-\frac{T_s^2(\mathbf{x}_{new})}{T_{c,s,\lim}^2}\right\}$$

$$P_{f,SPE}^{s,c}(\mathbf{x}_{new}|N) = \exp\left\{-\frac{SPE_s(\mathbf{x}_{new})}{SPE_{c,s,\lim}}\right\} \tag{9.17}$$

$$\begin{aligned} P^{s,c}_{f,T^2}(AB) &= P^{s,c}_{f,SPE}(AB) = \alpha \\ P^{s,c}_{f,T^2}(N) &= P^{s,c}_{f,SPE}(N) = 1 - \alpha \end{aligned} \tag{9.18}$$

where "N" represents normal, and "AB" represents abnormal. Then, the fault detection index (FDI) in each linear subspace can be constructed as

$$FDI^s_{T^2}(\mathbf{x}_{new}) = \sum_{c=1}^{C} [P^s_{T^2}(c|\mathbf{x}_{new}) P^{s,c}_{f,T^2}(AB|\mathbf{x}_{new})] \tag{9.19}$$

$$FDI^s_{SPE}(\mathbf{x}_{new}) = \sum_{c=1}^{C} [P^s_{SPE}(c|\mathbf{x}_{new}) P^{s,c}_{f,SPE}(AB|\mathbf{x}_{new})] \tag{9.20}$$

where $s = 1, 2, \ldots, k_c$. After the values of all fault detection indices have been calculated in different linear subspaces, a nonlinear fault detection index (NFDI) can be defined by combining these linear fault detection indices, thus

$$NFDI_{T^2}(\mathbf{x}_{new}) = comb\{FDI^s_{T^2}(\mathbf{x}_{new}), s = 1, 2, \ldots, k_c\} \tag{9.21}$$

$$NFDI_{SPE}(\mathbf{x}_{new}) = comb\{FDI^s_{SPE}(\mathbf{x}_{new}), s = 1, 2, \ldots, k_c\} \tag{9.22}$$

where $comb\{\cdot\}$ is the combination operator for the linear subspaces. For simplicity, the mean value of FDI in different linear subspaces can be used for nonlinear fault detection, which are given as follows

$$NFDI_{T^2}(\mathbf{x}_{new}) = mean\{FDI^s_{T^2}(\mathbf{x}_{new}),\ s = 1, 2, \ldots, k_c\} \tag{9.23}$$

$$NFDI_{SPE}(\mathbf{x}_{new}) = mean\{FDI^s_{SPE}(\mathbf{x}_{new}),\ s = 1, 2, \ldots, k_c\} \tag{9.24}$$

However, the mean value of FDI may be too conservative for nonlinear fault detection. In contrast, one can also use the maximum value of FDI to form the nonlinear fault detection indices, which are

$$NFDI_{T^2}(\mathbf{x}_{new}) = \max\{FDI^s_{T^2}(\mathbf{x}_{new}),\ s = 1, 2, \ldots, k_c\} \tag{9.25}$$

$$NFDI_{SPE}(\mathbf{x}_{new}) = \max\{\mathrm{FDI}^s_{SPE}(\mathbf{x}_{new}),\ s = 1, 2, \ldots, k_c\} \tag{9.26}$$

Compared to Eqs. (9.23) and (9.24), these two nonlinear fault detection indices may be too sensitive for fault detection, because any one violation of the linear subspaces can trigger the fault. On the opposite, if one constructs the nonlinear fault detection indices using the minimum value of FDI, the most conservative results may be obtained. Hence, a compromise can be made between these two extreme cases. The confidence limits of both $NFDI_{T^2}$ and $NFDI_{SPE}$ statistics are the same as the significance level α. Therefore, if $NFDI_{T^2}(\mathbf{x}_{new}) < \alpha$ and $NFDI_{SPE}(\mathbf{x}_{new}) < \alpha$, the process is judged to be normal. Otherwise, some fault is considered to be detected.

In summary, the new method can be outlined as the following steps:

Step 1: The normal process data under different operation modes are collected in database.

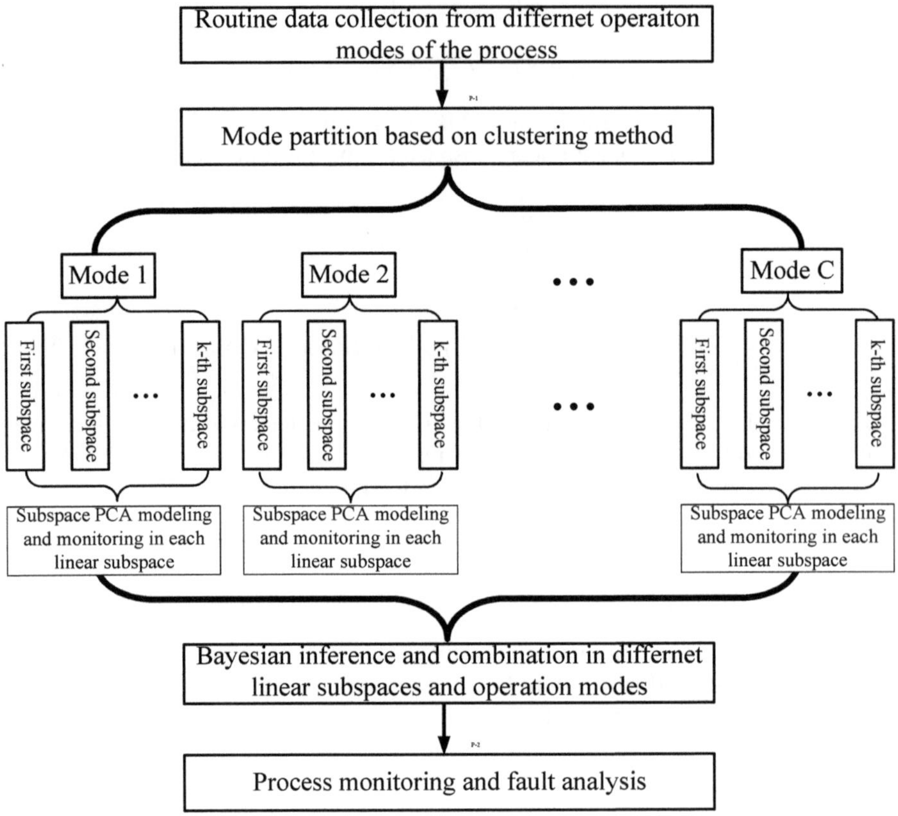

Fig. 9.2 Detailed description of the algorithm procedures

Step 2: Mode clustering and linear subspace construction by the methods is described in previous section.

Step 3: Develop linear subspace PCA models for different operation modes.

Step 4: For online monitoring of a new data sample, monitoring results in different operation modes are first generated in each linear subspace, and then combined by the Bayesian inference strategy.

Step 5: Combine FDI in different linear subspaces to form the final NFDI for nonlinear fault detection.

Step 6: If both the $NFDI_{T^2}$ and $NFDI_{SPE}$ statistic values are inside their corresponding confidence limits, this data sample is considered to be normal. If any of the two statistic values exceeds its confidence limit, fault is triggered and the process is considered to be abnormal.

The procedure of the algorithm is illustrated as in Fig. 9.2.

9.3.2 Algorithm Complexity Analysis

In this section, both offline and online computation complexities are analyzed. First, through analysis of the two-dimensional Bayesian (TDB) method in the offline phase, it can be found that the main time is spent in computing the covariance matrix for high-dimensional data. Therefore, we only focus on time complexity of covariance matrices as being computed by TDB. The time complexity to calculate the $m \times m$ covariance matrix of the original dataset $\mathbf{X} \in R^{n \times m}$ by PCA is given as follows

$$T_{PCA} = O(nm^2) \tag{9.27}$$

where $n > m$ always holds in most processes. Now, suppose we have partitioned the process into different operation modes and linear subspaces. The m_s' variables were selected in each linear subspace, and n_c data samples were assigned to each operation mode. Consider the time complexity of the constructed subspace PCA models, which can be calculated as

$$T_{SPCA} = O\left[\sum_{c=1}^{C} k_c n_c m_s'^2\right] \tag{9.28}$$

where k_c is the number of principal components selected in each operation mode, which is also the number of constructed linear subspaces. Taking the variable selection step into consideration, the total computation complexity of the TDB method can be calculated as

$$T_{TDB} = O\left(\sum_{c=1}^{C} n_c m^2\right) + O\left[\sum_{c=1}^{C} k_c n_c m_s'^2\right] \tag{9.29}$$

In contrast, the algorithm complexity of kernel PCA is $O(n^3)$. Basically, $n \gg m$ always holds in conventional industrial processes, and $\sum_{c=1}^{C} n_c = n$. Therefore, the following inequality also holds

$$O\left(\sum_{c=1}^{C} n_c m^2\right) + O\left[\sum_{c=1}^{C} k_c n_c m_s'^2\right] \ll O(n^3) \tag{9.30}$$

The online implementation complexity of the TDB method can be analyzed as follows. First, the complexity of the PCA method can be easily obtained as

$$TO_{PCA} = mk + m + 2k \tag{9.31}$$

Then, the online complexity of the new method can be given as

$$TO_{TDB} = \sum_{c=1}^{C} k_c (m_s' k_s + m_s' + 2k_s) \tag{9.32}$$

where k_c is the selected number of principal components for each operation mode, k_s is the selected number of principal components in each subspace, denote $\beta = m'_s/k_s$, Eq. (9.32) can be represented as

$$TO_{TDB} = O\left(\sum_{c=1}^{C} k_c \beta k_s^2\right) \tag{9.33}$$

In contrast, the online implementation of the traditional KPCA method is much more difficult, the complexity of which can be given as

$$TO_{KPCA} = 2mkn + n + 2k \tag{9.34}$$

Similarly, if we use the relationship $\gamma = m/k$, Eq. (9.34) becomes

$$TO_{KPCA} = O(n\gamma k^2) \tag{9.35}$$

It is clear that the online complexity of KPCA is much higher than that of TDB, because the values of β and γ are comparable, k is always greater than k_s, and $n \gg \sum_{c=1}^{C} k_c$ holds.

9.3.3 *Remarks*

Based on the last two subsections, the characteristics of the TDB method can be summarized as follows. First, the nonlinear and multimode behaviors of the process can be addressed simultaneously. Second, this method is unsupervised, little process or expert knowledge is needed. Third, the construction of linear subspaces for nonlinear approximation can greatly reduce the computational complexity for nonlinear modeling and monitoring. Fourth, the presentation of process monitoring results through a probabilistic framework can improve the comprehension of the process in both normal and faulty conditions. Besides, it is worth noticing that instead of using the Gaussian mixture model, two probability transformations have been made upon the traditional T^2 and *SPE* monitoring statistics. This is because the computation of the Gaussian mixture model is much more complex, and it may also cause singular problem. On the other hand, through these two transformations, the method implementation becomes easier, and the singular problem can also be avoided. Furthermore, depending on this model structure, extensions of this method to other monitoring statistics are straightforward. However, it is much more difficult to extend the Gaussian mixture model to other cases.

9.4 Case Study of TE Benchmark Process

In this section, the performance of the TDB method is evaluated through the Tennessee Eastman process. Similarly, 16 continuous process variables which are the

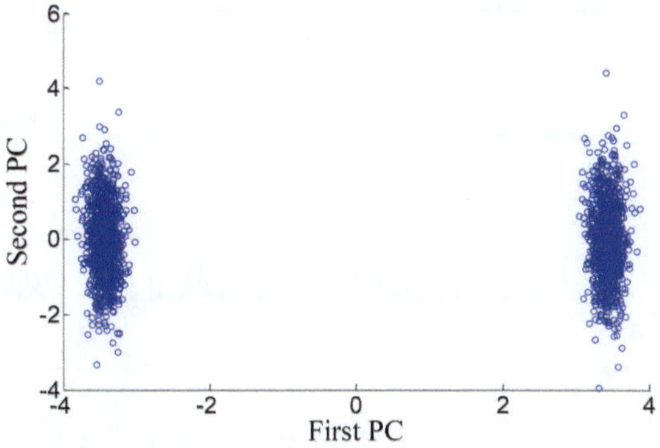

Fig. 9.3 Scatter plot of the first two principal components for the process data

same as several previous chapters are selected for process monitoring. There are six operation modes available for simulation case studies, two of which are selected for the simulation of multimode processes. The simulation datasets which we have collected were separated into two parts: the training datasets and the testing datasets, both of which consisted of 1,000 observations. Their sampling interval was 3.6 min. All process faults were introduced after sample 500.

To develop the TDB monitoring model, the process data are first partitioned into different operation modes. By examining the score plot of the first two principal components in Fig. 9.3, we can find that the process is run under two different operation conditions. Then, PCA decomposition is carried out upon each operation mode, which determines the subspace direction for the corresponding operation mode. The number of principal components in each operation mode is selected by the cumulative percentage variance (CPV) rule, thus the explained variance of data information is larger than 85 %. As a result, eight variables are selected in each linear subspace in the first variable selection step, and six of them are finally determined for linear subspace construction. The initial number of variables in each linear subspace can be selected as 1/2∼1/3 of the total number of the process variable, which also depends on the number of linear subspaces. When a small number of linear subspaces has been constructed, the number of variables in each subspace should be selected as a relatively big value, in this case the combination of different linear subspaces could overlap the whole nonlinear process. Next, subspace PCA models are built in these constructed linear subspaces. Similarly, the number of principal components in each subspace PCA model is determined by the CPV rule, the value of which is also selected as 85 %. The confidence level of each monitoring model is selected as 99 %. For simplicity, the nonlinear fault detection indices are constructed through Eqs. (9.23) and (9.24). In order to compare the monitoring performance of the TDB method to other methods, the traditional PCA and kernel PCA models are also developed. The principal component numbers for both methods are also determined by the CPV rule, with similar data variances explained by them.

To evaluate the feasibility of the TDB method, a normal dataset is first tested, which is collected from the two operation modes, with 500 samples each. Monitoring

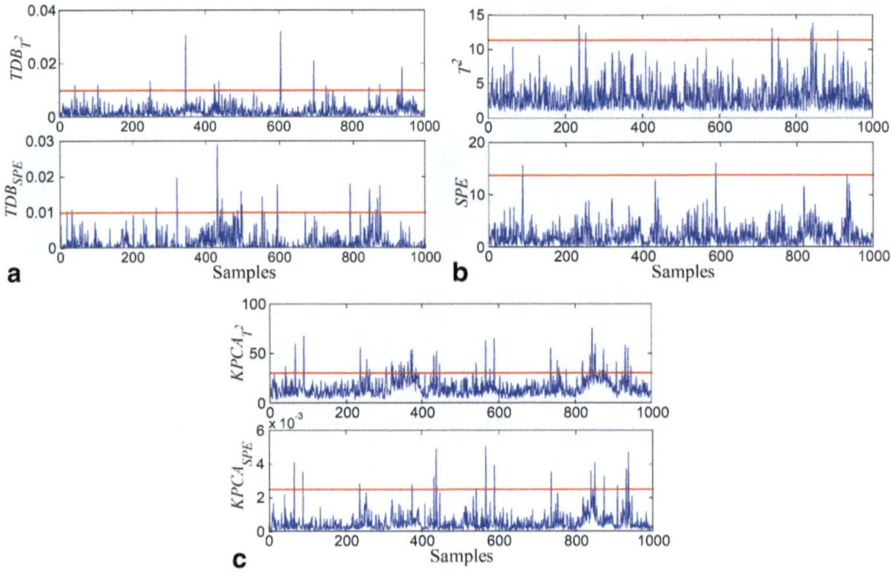

Fig. 9.4 Monitoring results of normal process. **a** TDB. **b** PCA. **c** KPCA

Table 9.1 False alarm rates of monitoring statistics

Monitoring statistics	False alarm rates (%)
TDB NFDI_T^2	1.3
TDB NFDI_SPE	2.7
PCA T^2	0.7
PCA SPE	0.2
KPCA T^2	6.1
KPCA SPE	1.6

results are given in Fig. 9.4a, which shows that no fault has been triggered by the new method. For comparison, the monitoring results of PCA and KPCA are also given, which are shown in Fig. 9.4b, c. As we know, the false alarm rate is a key index for feasibility evaluation of the monitoring method. Table 9.1 shows false alarm rates of all monitoring statistics of different methods. It can be found from the table that the KPCA method has the largest false alarm rate among all the three methods. In contrast, the false alarm rates of both monitoring statistics of the TDB method are acceptable, and the PCA method appears to be the "best." However, if we monitor this normal dataset with PCA subspace monitoring models, which were built separately in each operation mode, the monitoring results are totally different. These subspace monitoring results for the normal dataset are given in Fig. 9.5. One can find that the first 500 samples of the process are judged to be normal by the first subspace model, while the last 500 samples are considered normal by the second subspace model. This is because the first 500 samples were collected under the first operation mode, and the last 500 samples were collected under the second one. When the first 500

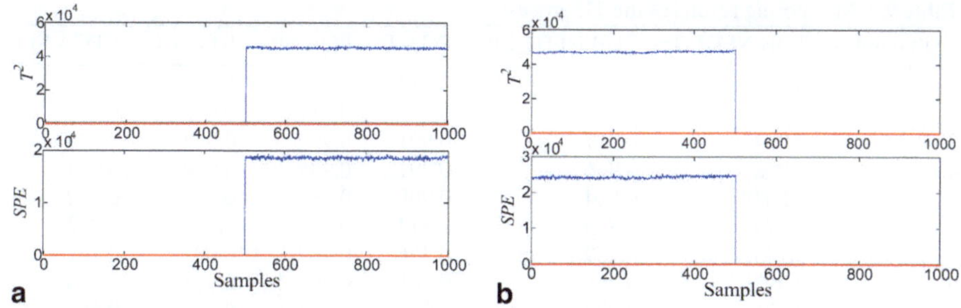

Fig. 9.5 Monitoring results of normal process. **a** First subspace. **b** Second subspace

samples are monitored by the second subspace model, they are treated as faults, as no information of the first operation mode was contained and trained in the second subspace model.

The situation of the last 500 data samples is similar. However, our new method can efficiently integrate monitoring results in different operation mode. In another word, it is not necessary to select and change subspace models for monitoring different data samples. Although we can use different operation mode datasets to build a global statistical monitoring model, the determined confidence limits may be too conservative and loose, which can deteriorate the fault detection performance.

The superiority of the TDB method to the traditional PCA and KPCA methods is illustrated through evaluations of the process faults. It is noted that in this simulation study, the process will be shut down when fault 6 was introduced. Therefore, fault 6 was excluded from the present case study, thus a total of 19 faults are evaluated. Mean values of 100 monitoring results (miss detection rates) of these 19 process faults by all three methods are summarized in Table 9.2. Good monitoring performance is considered to be obtained when the miss detection rate is small. Hence, the smaller of the miss detection rate, the better monitoring performance of the method shows. For each process fault, the smallest miss detection rate is marked with a star. One can find that the best monitoring results are obtained by the TDB method and KPCA for most of the process faults. Particularly, compared to the traditional PCA method, the performances of faults 2, 10, 11, 14, and 17 have been greatly improved by the new method. Although KPCA performs a little better than TDB for faults 1, 2, 7, and 8, their performances are comparative, which are both much better than that of PCA. However, the monitoring results of TDB for faults 5, 10, 11, and 14 are much better than those of KPCA. In summary, average miss detection rates of all monitoring statistics for these process faults are given in the last row of Table 9.2. Obviously, two monitoring statistics of the TDB method give the best results. More clearly, monitoring results of two process faults (faults 11 and 14) are illustrated in Figs. 9.6 and 9.7, respectively. As mentioned above, each faulty dataset contains 1,000 data samples, among which the last 500 are abnormal. The faulty dataset can be generated upon the first operation mode or the second one. In this simulation, we put these two faulty datasets together, thus a total number of 2,000 samples are generated, among

Table 9.2 Monitoring results of the TE process

Fault number	TDB NFDI_T^2	TDB NFDI_SPE	PCA T^2	PCA SPE	KPCA T^2	KPCA SPE
1	0.033	0.038	0.314	0.070	0.028*	0.038
2	0.079	0.093	0.811	0.617	0.045*	0.225
3	0.994	0.997	0.991*	0.999	0.934	0.985
4	0.968	0.966	0.991	0.998	0.928*	0.974
5	0.503*	0.504	0.968	0.987	0.860	0.892
7	0.874	0.825	0.914	0.909	0.759*	0.852
8	0.143	0.159	0.356	0.467	0.091*	0.141
9	0.993	0.968	0.985	0.998	0.911*	0.969
10	0.121	0.073*	0.948	0.974	0.808	0.589
11	0.132*	0.138	0.846	0.882	0.564	0.464
12	0.459	0.510	0.496	0.538	0.350*	0.436
13	0.147	0.262	0.290	0.376	0.083*	0.207
14	0.030*	0.040	0.988	0.995	0.874	0.321
15	0.995	1.000	0.993	0.999	0.940*	0.988
16	0.995	0.998	0.993	0.999	0.940*	0.987
17	0.048	0.037*	0.860	0.358	0.238	0.091
18	0.549*	0.558	0.760	0.845	0.582	0.628
19	0.544	0.497*	0.774	0.859	0.578	0.664
20	0.209	0.145*	0.545	0.680	0.267	0.276
Average	0.464*	0.464*	0.780	0.766	0.567	0.565

*The best result among all monitoring statistics

which samples 1–500 and 1,001–1,500 were collected under the first and second normal operation mode, respectively. Faults 11 and 14 were introduced upon these two operation mode, thus samples 501–1,000 and 1,501–2,000 were collected. By examining and comparing monitoring results of the three methods in Figs. 9.6 and 9.7, the superiority of the TDB method is obvious, as the monitoring performance of these two process faults have been greatly improved.

Next, we intend to examine the selection strategy for result combination of the linear subspaces. For simplicity, we have used the mean FDI value of linear subspaces to form NFDI. It was demonstrated that this combination strategy has gained good monitoring results for both normal and faulty process datasets. If we select the maximum FDI value among the constructed linear subspaces, the TDB method is expected to be very sensitive for fault detection. However, false alarm rates of both monitoring statistics may also arise. On the other hand, if we select the minimum FDI value of linear subspaces, false alarm rates can be reduced, but at the same time the miss detection rates may arise, which means the fault detection sensitivity is also reduced. According to the CPV rule, we have selected 12 principal components for linear subspace construction. Therefore, except for the selection of the mean value, we have another 12 choices, which are between the two extreme cases described in previous section. Denote these 12 combination cases from the maximum FDI value to the minimum one as case 1 to case 12, a series of monitoring results can be obtained, including the false alarm rate and the miss detection rate.The result of false alarm rates for the normal dataset and average miss detection rates of all 19 faulty datasets are shown together in Figs.9.8 and 9.9. As expected, when the combination strategy

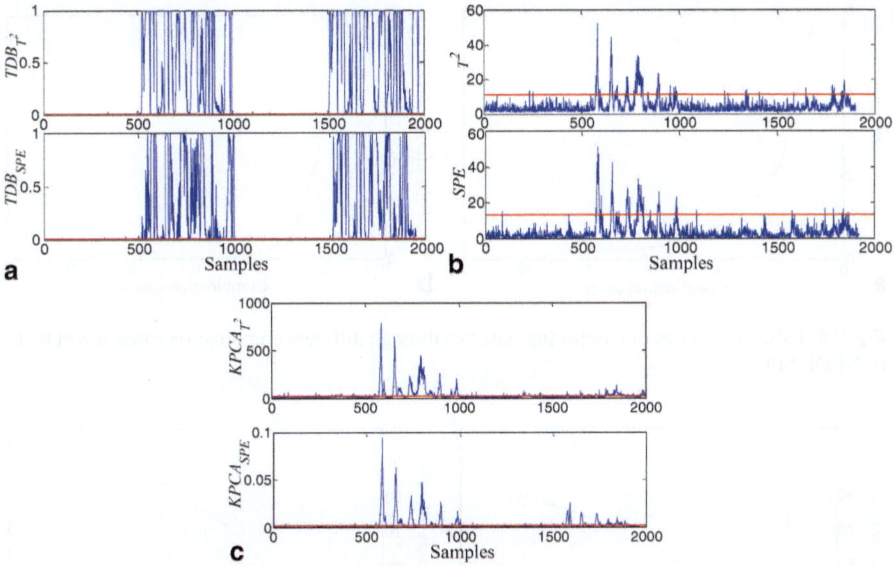

Fig. 9.6 Monitoring results of fault 11. **a** TDB. **b** PCA. **c** KPCA

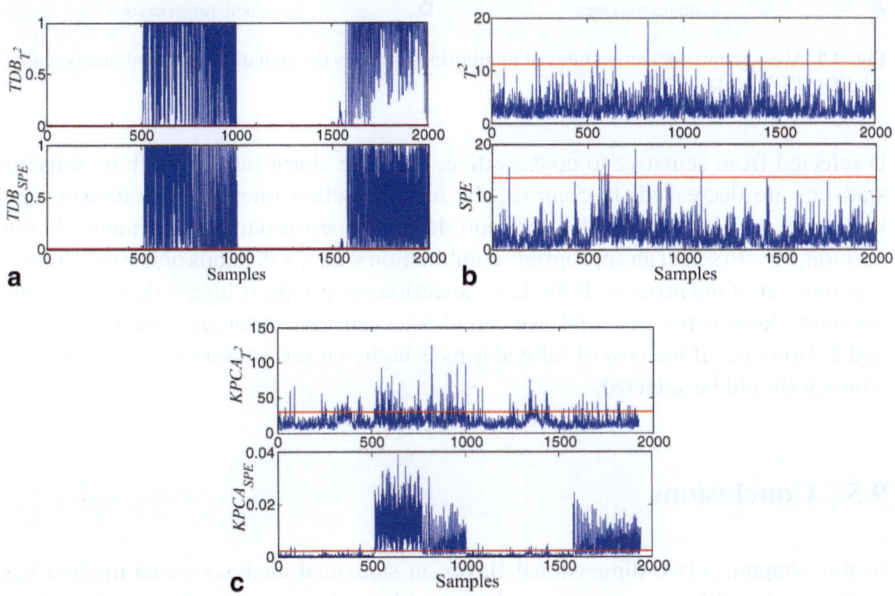

Fig. 9.7 Monitoring results of fault 14. **a** TDB. **b** PCA. **c** KPCA

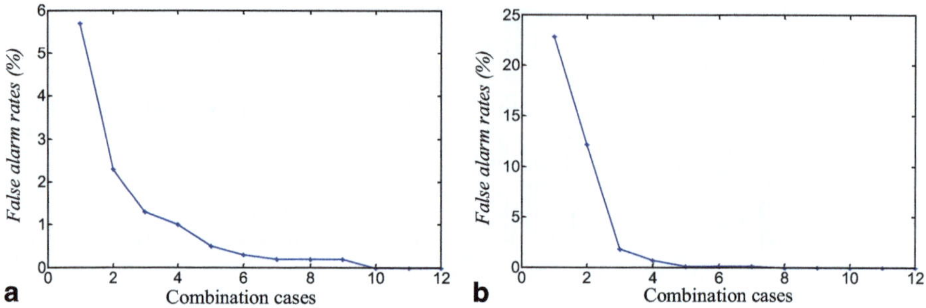

Fig. 9.8 False alarm rates of monitoring statistics through different combination cases. **a** NFDI_T^2. **b**) NFDI_SPE

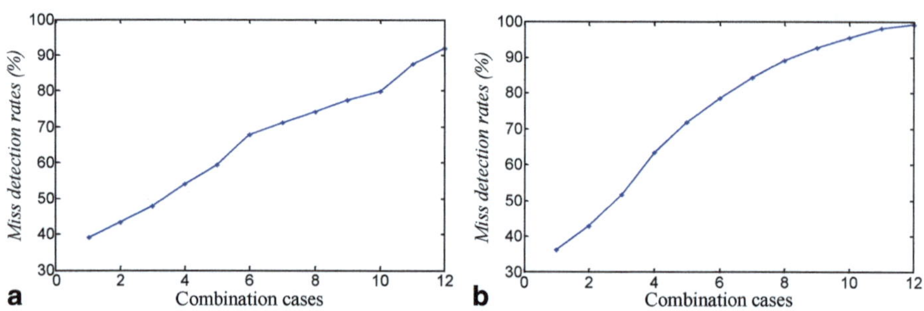

Fig. 9.9 Average miss detection rates of monitoring statistics through different combination cases. **a** NFDI_T^2. **b** NFDI_SPE

is selected from sensitive to conservative, the false alarm rates of both monitoring statistics are decreased. In contrast, the miss detection rates of both statistics are increased, which indicates that the fault detection performance is degraded. In our opinion, how to select an appropriate combination strategy depends on the monitoring requirement of the process. If the fault detection sensitivity is highly demanded, and the false alarm is not essential, we can choose sensitive strategies, such as cases 1 and 2. However, if the cost of false alarms is high, a more conservative combination strategy should be selected.

9.5 Conclusions

In this chapter, a two-dimensional Bayesian statistical analysis-based method has been introduced for monitoring nonlinear multimode processes. In order to reduce the computation complexity for nonlinear process monitoring, the concept of linear subspace approximation was introduced. A two-step variable selection strategy was developed for linear subspace construction, which can simultaneously guarantee diversity and linearity of the linear subspace. Bayesian inference was introduced to integrate monitoring results obtained in different operation modes. To facilitate subspace monitoring result combination, the monitoring task in each linear subspace was implemented through a probabilistic manner. The feasibility and efficiency of the TDB method were evaluated in TE benchmark process. Compared to other monitoring methods, encouraging results have been obtained by the new method.

When monitoring for a new data sample, we may want to know which operation mode the process is operating on. In this case, mode localization becomes more important, as it can improve the understanding of the process for the engineer. Through the Bayesian inference, the new method has its potential for mode localization and fault identification. Using the posterior probability that has been obtained through Bayesian inference, the current operation mode for the monitored data sample can be easily located. Particularly, if the monitored sample comes from the transition between two operation modes, this data sample will be located into both operation modes with different posterior probability values. Another important potential of this method is for fault identification, which can be regarded as a pattern match problem. If we build a monitoring model for each fault case through this method, a fault pool can be constructed. When some fault has been detected through the monitoring approach, it can be identified by the posterior probabilities, which can be calculated through Bayesian inference among previous fault cases.

Chapter 10
Dynamic Process Monitoring

10.1 Introduction

The early detection of incipient fault conditions is of fundamental importance for the operation of industrial systems, such as mechanical systems, chemical systems, biological systems, etc. The consequences of not being able to detect such faults at early stages can, for example, include reduced productivity in manufacturing processes, reduced efficiency of engines, equipment damage or even failure. Therefore, early detection of such faults can provide significant improvements in the reduction of operational and maintenance costs, system down-time and lead to increased levels of safety, which is of ever-growing importance. For example, an incipiently developing fault in a mechanical system usually affects certain parameters, such as vibration, noise and temperature. Therefore, the analysis of these "external variables" allows the monitoring of internal components, such as gears, which are usually inaccessible without the dismantling of the system. It is consequently essential to extract relevant information from the recorded signals for the aim of detecting any irregularities that could be caused by such faults.

This chapter takes a gearbox as an example to demonstrate the importance of carrying our dynamic process monitoring in industrial systems. In the gearbox system, if running speed and loading conditions are kept constant, the vibration signals are influenced by defects, such as deformation, breakage or fracture of the teeth. Over the last few decades, various techniques have been proposed, including averaged vibration signal, spectrum and time-frequency techniques, as well as amplitude and phase techniques. Recent work in these areas are done by Staszewski and Tomlinson 1997; Baydar and Ball 2001; Stander et al. 2002; Tan and Mba 2005; Hu et al. 2007; Wu et al. 2008; Wang 2008; Eftekharnejad and Mba 2008; Vincenzo et al. 2008; Bartelmus and Zimroz 2009. Due to the conceptual simplicity, Fugate et al. (2001) recently showed that applying statistical process control technique in conjunction with autoregressive filters could detect a progressive damage of a concrete bridge. As vibration information may be available from more than one sensor, Baydar et al. (1999, 2001) introduced the application of multivariate statistical process control (MSPC) to successfully detect tooth defects. In a comprehensive comparison, Baydar et al. (2001) showed that multivariate statistical techniques are equally as sensitive in detecting

Z. Ge, Z. Song, *Multivariate Statistical Process Control,* Advances in Industrial Control,
DOI 10.1007/978-1-4471-4513-4_10,

tooth defects as signal-based approaches but easier to implement in practice. More recent applications of MSPC methods in monitoring mechanical systems have been reported (Malhi and Gao 2004; He et al. 2007a, 2009).

Despite the reported success of MSPC-based work in detecting faults by analyzing recorded vibration signals, the research literature has not considered data structures that can represent correlation among and between these signals. Moreover, the MSPC monitoring scheme assumes that the vibration signals follow a Gaussian distribution. However, the analysis of experimental data in the gearbox system shows that readings from vibration sensors may not possess such statistical properties. To address these issues, this chapter introduces two approaches that can monitor correlated vibration signals which are non-Gaussian. Proposed work on dynamic MSPC (Ku et al. 1995; Xie et al. 2006) showed that an incorrect data model can affect the number of type I and type II errors, whilst the same effect is noticeable for assuming an incorrect distribution function for the vibration signals (He et al. 2007b; Widodo et al. 2007; Widodo and Yang 2007).

Although a number of research studies have been carried out on the dynamic and non-Gaussian data behaviours separately, such as dynamic principal component analysis (DPCA) (Ku et al. 1995; Xie et al. 2006; Chen and Liu 2002), improved DPCA (Kruger et al. 2004), subspace model identification (SMI) (Xie et al. 2006; Overschee and Moor 1996; Li and Qin 2001) for the dynamic issue, independent component analysis (ICA) (Bishop 1999; Breiman 1996; Chen and Liu 1999), support vector machine (Widodo and Yang 2007), support vector data description (SVDD) (Liu et al. 2008; Ge et al. 2009b) for the non-Gaussian issue, the research literature has not offered a comprehensive treatment that considered both dynamic and non-Gaussian behaviour.

10.2 Two Dynamical Process Monitoring Methods

10.2.1 First Method

The first technique is based on a dynamic ICA (DICA) formulation to produce source signals that (Lee et al. 2004), in turn, generate correlated and non-Gaussian vibration signals. Utilizing the SVDD method then allows performing a statistical inference to determine whether the recorded signals correspond to an in-statistical or out-of-statistical-control situation. With the latter scenario being indicative of a fault condition, this method represents a dynamic extension of the recent work by Liu et al. (2008).

10.2.2 Second Method

The second algorithm relies on the utilization of the statistical local approach (LA) to monitor changes in parameters for computing the state sequences (Basseville and Nikiforov 1993; Basseville 1998). The underlying data model, describing the correlated vibration signals, is of Kalman innovation form and represents a state space model. The statistical LA has been proposed as a method for detecting abrupt changes (Basseville and Nikiforov 1993), and overcomes the effects of non-Gaussian process data (Kruger and Dimitriadis 2008). This technique has been applied in many areas, such as bridge damage detection and mechanical vibration monitoring (Basseville 1998; Zhang et al. 1994; Basseville et al. 2000). A recent study by Kruger et al. (2007) proposed the integration of statistical local in an MSPC framework to detect incipient changes in the variable covariance structure. This second approach incorporates the statistical LA into the dynamic monitoring framework which is based on SMI and, therefore, represents a dynamic extension of the work by Kruger et al. (2007) and Kruger and Dimitriadis (2008).

10.3 Illustration of the Gearbox System

This section describes the specification of the gearbox system used in this chapter and discusses how fault conditions of various magnitudes were injected into this system. This is an important research area, given that a gearbox is an arrangement involving a train of gears by which power is transmitted from the engine to the axle of a car, for example.

10.3.1 Description of the Gearbox System

A schematic diagram of the two-stage helical gearbox system is given Fig. 10.1. A tooth breakage in a gearbox, which is a serious localized fault, was simulated by removing a certain percentage of one tooth in the pinion gear. This tooth removal enables the experimental representation of gradual fault advancement to test the statistical-based fault detection schemes developed here. Each experiment was performed under the full-load condition of 260 N m. Further specifications of the gearbox are summarized in Table 10.1.

Similar to previous research (Baydar and Ball 2001; Baydar et al. 2001), which studied the same problem, four accelerometers recorded the vibration of the gearbox under full-load conditions. The vibration signals were simultaneously sampled at a frequency of 6.4 Hz. Each recorded dataset contained 2,620 samples, which corresponded to 7.81 revolutions of the pinion.

10.3.2 Analysis of Reference Data

Figure 10.2 shows plots of the first 1,000 samples per each of the four sensor readings describing the reference conditions (no fault).

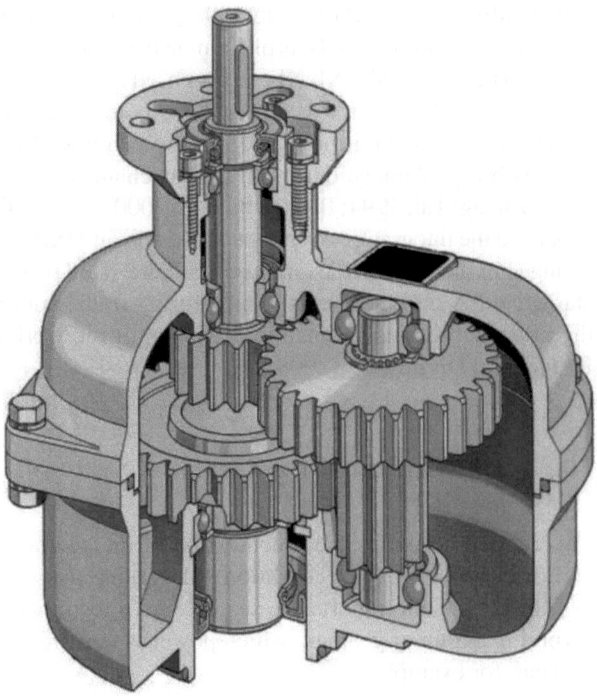

Fig. 10.1 Schematic diagramand sketch of the of gearbox system used in this study

Table 10.1 Specification of the gearbox system

Specification	Number of teeth	Speed (rev/s)	Meshing frequency (Hz)	Contact ratio	Overlap ratio
1st stage	34/70	24.33	827.73	1.359	2.89
2nd stage	29/52	6.59	342.73	1.479	1.479

Figure 10.3 depicts the auto- and cross-correlation functions constructed from these reference samples.

By inspection, significant auto- and cross-correlation can be observed. This is expected since the vibrating gearbox represents a dynamic system and any imbalance will consequently result in a dynamic response. The correlation, in turn, implies that employing a static data model of the form

$$\mathbf{z}(k) = \Xi\,\mathbf{s}(k) + \mathbf{e}(k), \tag{10.1}$$

where $\mathbf{z}(k) \in R^N$ is the vector storing the sensor reading of the kth instance, $\Xi \in R^{N \times n}$, $n < N$ is a parameter matrix, $\mathbf{s}(k) \in R^n$ represents the source signals and $\boldsymbol{e}(k) \sim N\left\{0, \sigma_e^2 \mathbf{I}\right\}$ $\forall k$ is a statistically independent measurement error vector, is not suitable, although, a common assumption in MSPC. A further assumption that is often imposed on the source and error variables relates to their first and second order properties

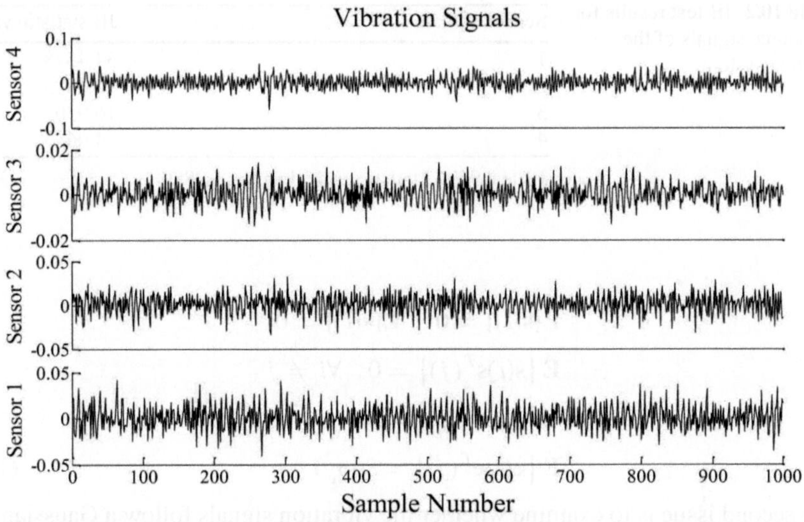

Fig. 10.2 Signals of vibration sensors. (Reference condition)

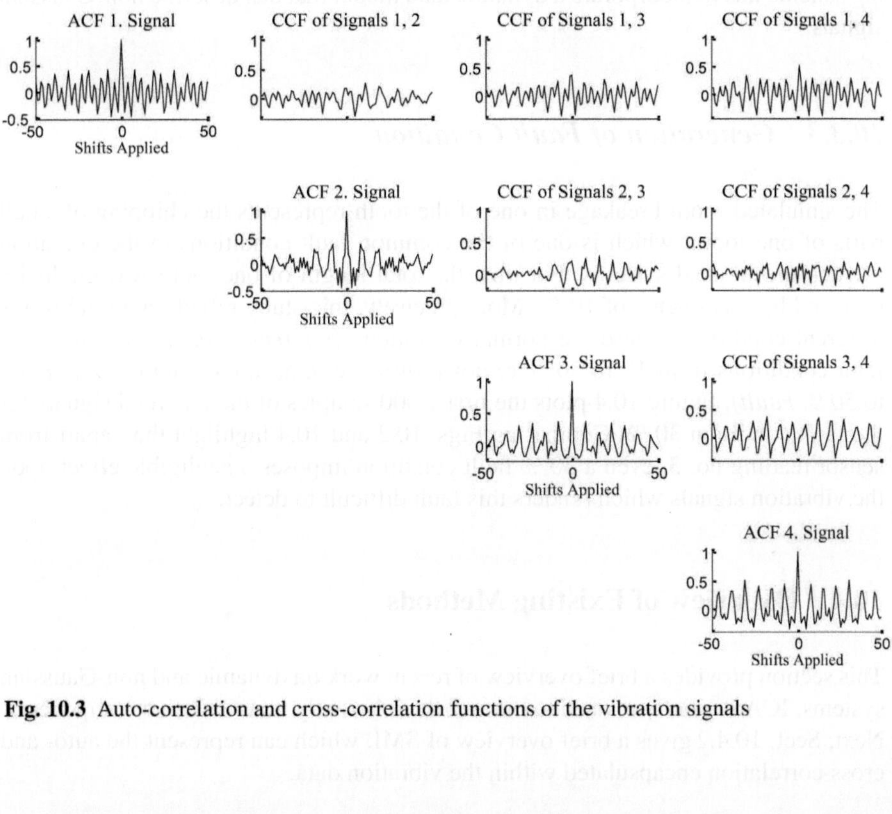

Fig. 10.3 Auto-correlation and cross-correlation functions of the vibration signals

Table 10.2 JB test results for the source signals of the gearbox system

Sensor signals	JB statistic values
1	34.4578
2	74.8885
3	16.9303
4	4.0683

JB test 95 % statistic critical value: 5.9880

$$\begin{aligned} &\mathbf{E}\{\mathbf{s}(i)\} = 0 \quad \mathbf{E}\{e(i)\} = 0 \\ &\mathbf{E}\left\{\mathbf{s}(i)\mathbf{s}^T(j)\right\} = 0 \quad \forall i \neq j \\ &\mathbf{E}\left\{\mathbf{e}(i)\mathbf{s}^T(j)\right\} = 0 \quad \forall i, j \in N \\ &\mathbf{E}\left\{\mathbf{e}(i)\mathbf{e}^T(j)\right\} = \delta_{ij}\sigma_e^2\mathbf{I}. \end{aligned} \tag{10.2}$$

The second issue is to examine whether the vibration signals follow a Gaussian distribution. Table 10.2 shows the results of the well-known JB test (Jarque and Bera 1987) and indicates that the first three vibration signals in the reference data are highly non-Gaussian. Therefore, the analysis shows that a statistical-based monitoring scheme has to incorporate a dynamic data model that can describe non-Gaussian signals.

10.3.3 Generation of Fault Condition

The simulated tooth breakage in one of the tooth represents the chipping of small parts of one tooth, which is one of the common fault conditions in the operation of such mechanical systems. For this, the total length of one tooth was gradually removed by increments of 10 %. More precisely, this study relied on a total of six different conditions, these are normal conditions (*reference conditions*), and five fault conditions from 10 to 50 % removal over the total tooth length (*10 % Fault* to *50 % Fault*). Figure 10.4 plots the first 1,000 samples of the recorded signals for the fault condition 30 %. Comparing Figs. 10.2 and 10.4 highlight that, apart from sensor reading no. 3, even a 30 % fault condition imposes a negligible effect upon the vibration signals which renders this fault difficult to detect.

10.4 Overview of Existing Methods

This section provides a brief overview of recent work on dynamic and non-Gaussian systems. ICA is briefly revised to unravel the inherently non-Gaussian components. Next, Sect. 10.4.2 gives a brief overview of SMI, which can represent the auto- and cross-correlation encapsulated within the vibration data.

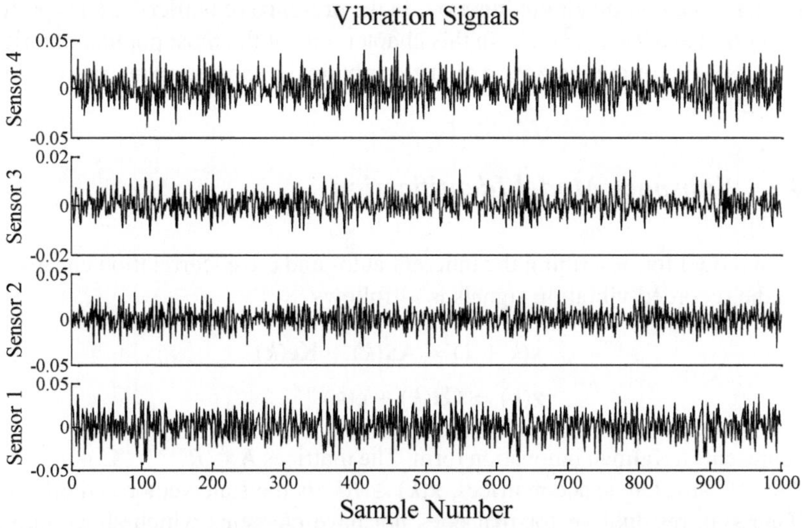

Fig. 10.4 Signals of the vibration sensors (30 % fault)

10.4.1 Independent Component Analysis

Given that the vibration signals from a mechanical system may not be assumed to follow a Gaussian distribution, ICA is applied to obtain independent components (ICs) from the extracted vibration signals. The ICA model assumes the existence of independent source variables $\mathbf{s}(k) \in R^n$ that produce a variable $\mathbf{z}(k)$ which is shown in Eq. (10.1). The aim of ICA is to find a separate matrix $\mathbf{W}$, such that

$$\hat{\mathrm{s}}(k) = \mathbf{W}\mathbf{z}(k) = \mathbf{W}\,\Xi\mathbf{s}(k) \approx \mathbf{s}(k). \tag{10.3}$$

Including a whitening procedure, $\mathbf{y}(k) = \mathbf{Q}\mathbf{z}(k), \mathbf{Q} \in R^{N\times N}$ being the whitening matrix, the separating system matrix can be rewritten as follows

$$\hat{\mathrm{s}}(k) = \mathbf{W}\mathbf{z}(k) = \mathbf{W}\mathbf{Q}^{+}\mathbf{y}(k) = \mathbf{B}^{T}\mathbf{y}(k), \tag{10.4}$$

where $[\cdot]^{+}$ is a generalized inverse and $\mathbf{B} \in R^{N\times n}$ is determined to maximize the non-Gaussianity of $\hat{\mathrm{s}} = \mathbf{B}^{T}\mathbf{y}$ under the constraint that the columns of $\mathbf{B}$ are mutually orthogonal and determined by maximizing $J(y)$, $\mathrm{y}=\mathbf{b}^{T}\mathbf{x}$, which measures non-Gaussianity. For $J(y)$, the negentropy is usually employed, which relies on the information-theoretic quantity of differential entropy, defined as $H(y) = -\int f(y)\log[f(y)]dy$, where y and $f(\cdot)$ is a random variable and its density function. A Gaussian variable v has the largest entropy among all random variables of equal variance, and allows the definition of $J(y) = H(v) - H(y)$, which can be approximated by

$$J(y) \approx [E\{G(y)\} = E\{G(v)\}]^{2}. \tag{10.5}$$

Here, $G(\cdot)$ is the non-quadratic function. In the presence of outliers, $G(\cdot)$ is preferred to be $-(1/a_2)\exp[(-a_2 y^2)/2\ ]$. In this chapter, one of the most popular algorithms, FastICA, is used for maximization of $J(y)$ to obtain the separating matrix (Hyvarinen and Oja 2000).

10.4.2 Subspace Model Identification

The data model for describing the inherent auto- and cross-correlation encapsulated within the recorded vibration signals is as follows

$$\mathbf{x}(k+1) = \mathbf{A}\mathbf{x}(k) + \mathbf{K}\mathbf{e}(k) \tag{10.6}$$

$$\mathbf{z}(k) = \mathbf{C}\mathbf{x}(k) + \mathbf{e}(k)$$

and represents a Kalman innovation form. The matrices $\mathbf{A} \in R^{n_x \times n_x}$, $\mathbf{C} \in R^{N \times n_x}$ and $\mathbf{K} \in R^{n_x \times N}$ are state space matrices, $\mathbf{x}(k) \in R^{n_x}$ is the state vector and $\mathbf{e}(k) \in R^{n_x}$ is a Gaussian residual vector that does not have elements which show auto- and cross-correlation.

The application of a **QR** decomposition to the block Hankel matrices storing the sensor readings to form the past and future arrangement $\mathbf{Z}_p$ and $\mathbf{Z}_f$, respectively, (Bauer et al. 1999)

$$\begin{bmatrix} \mathbf{Z}_p \\ \mathbf{Z}_f \end{bmatrix} = \begin{bmatrix} \mathbf{R}_{11} & \mathbf{0} \\ \mathbf{R}_{21} & \mathbf{R}_{22} \end{bmatrix} \begin{bmatrix} \mathbf{Q}_1 \\ \mathbf{Q}_2 \end{bmatrix}. \tag{10.7}$$

A subsequent singular value decomposition reveals the number of state sequence and estimates of **A**, **C**, **K** up to a similarity transformation

$$\mathbf{R}_{21} = [\mathbf{U}_1 \ \ \mathbf{U}_2] \begin{bmatrix} \mathbf{S}_1 & \mathbf{0} \\ \mathbf{0} & \mathbf{S}_2 \end{bmatrix} \begin{bmatrix} \mathbf{V}_1^T \\ \mathbf{V}_2^T \end{bmatrix}. \tag{10.8}$$

The indices "1" and "2" refer to the left and right singular vectors that correspond to the dominant and lower order singular values, which are stored in $\mathbf{S}_1$ and $\mathbf{S}_2$, respectively. The estimated state matrix $\mathbf{X}_f$ can be obtained as $\mathbf{X}_f \approx \Psi\mathbf{Z}_p = \Gamma^{+}\mathbf{R}_{22}\mathbf{R}_{11}^{-1}\mathbf{Z}_p$, where the "+" represents the generalized inverse and $\Gamma = \mathbf{U}_1\mathbf{S}_1^{1/2}$. Then, the Kalman innovation sequences $\mathbf{e}(k)$ can be formulated iteratively as $\mathbf{e}(k) = \mathbf{z}(k) - \mathbf{C}\mathbf{x}(k)$.

10.5 Dynamic Monitoring Schemes for Non-Gaussian Vibration Systems

This section utilizes existing work on dynamic modelling and the monitoring of non-Gaussian systems to develop two competitive techniques suitable for the recorded vibration signals. The first method, which is described in Sect. 10.5.1, incorporates a DICA formulation into the existing ICA-SVDD approach. Section 10.5.2 then proposes the joint application of a Kalman innovation model and the statistical LA.

10.5.1 DICA-SVDD Method

This monitoring method consists of two steps: a DICA formulation and a subsequent SVDD formulation to produce a univariate monitoring statistic (Tax and Duin 1999). The first step entails the dynamic arrangement of the original vibration signals

$$\mathbf{z}_d^T(k) = [\mathbf{z}^T(k-d)\ \mathbf{z}^T(k-d+1)\ \mathbf{z}^T(k-1)], \tag{10.9}$$

where the index "k" represents the current sample and "d" is the dynamic order of auto-regressive arrangement. Then the ICA algorithm is carried out upon the new data sample $\mathbf{z}_d(k) \in R^{Nd}$ such that the ICs are

$$\hat{\mathbf{s}}_d(k) = \mathbf{W}_d\mathbf{z}_d(k), \tag{10.10}$$

where $\mathbf{W}_d$ is the separating matrix of the "dynamic" ICA model. The residual vector can be calculated as

$$\mathbf{e}_d(k) = \mathbf{z}_d(k) - \mathbf{A}_d\hat{\mathbf{s}}_d(k), \tag{10.11}$$

where $\mathbf{A}_d$ is the mixing matrix, which can be determined as shown by Lee et al. (2004). Due to the complex distribution of the IC, SVDD is then employed for determining a statistical confidence limit, which according to Chap. 3 is equal to the squared radius of the hypersphere.

From the SVDD representation of the dynamic ICs, the univariate T^2 statistic can be defined as (Ge et al. 2010b)

$$T^2(k) = d^2\left\{\Phi[\hat{\mathbf{s}}_d(k)]\right\} = \left\|\Phi(\hat{\mathbf{s}}_d(k) - \mathbf{a}\right\|^2 \tag{10.12}$$

where $\mathbf{a}$ is the center of the hypersphere in the SVDD model, and the confidence limit of the T^2 statistic is equal to the radius of the hypersphere, R^2. A second univariate statistic can be constructed from the residuals of the ICA model

$$Q(k) = \mathbf{e}_d^T(k)\mathbf{e}_d(k). \tag{10.13}$$

By approximating the Q statistic using a χ^2 distribution, the confidence limit $Q\alpha$ for a significance α is given by the product $g\chi_h^2$, where g and h are

$$\begin{aligned} gh &= E\{Q\} = \mu_Q \\ 2g^2h &= E\left\{(Q-\mu_Q)^2\right\} = \sigma_Q^2, \end{aligned} \tag{10.14}$$

where μ_Q and σ_Q^2 are the mean and variance of Q, respectively.

10.5.2 SMI Technique Using the Statistical LA (SMILA)

For the introduction of this method, we first outline the modelling of parameter changes of the dynamic data model, which is followed by a brief summary of the statistical LA prior to the generation of improved residuals. Finally, univariate monitoring statistics are derived to conduct the monitoring task.

In order to model the changes in the model parameter, we can use the computation of the state sequences

$$\mathbf{x}(k) = \Psi \mathbf{z}_p(k), \tag{10.15}$$

where $\mathbf{z}_p(k)$ represents the sequence in the "past" Hankel matrix. Denoting the original Ψ matrix as Ψ_0, which is estimated by analyzing the entire reference dataset, the incorporation of the LA relies on a moving window approach. The samples within an ith window produce the Ψ matrix Ψ_i, which allows the construction of a primary residual for the LA

$$\mathbf{\Phi}(\boldsymbol{\theta}, i) = (\Psi_0 - \Psi_i)\mathbf{z}_p, \tag{10.16}$$

where $\mathbf{z}_p^T = [\mathbf{z}^T(k-s) \;\; \mathbf{z}^T(k-s+1) \;\; \mathbf{z}^T(k-1)]$ represents the sequence of the past Hankel matrix in the moving window, and $\boldsymbol{\theta}$ is the parameter set of the system. However, the use of subspace identification only allows the state sequences to be computed up to a similarity transformation. To guarantee that the primary residuals are not dependent on an arbitrary similarity transformation, the singular values of the matrices Ψ_0 and Ψ_i, which are invariant to a similarity transformation, can be utilized. This allows the following construction of the primary residuals (Ge et al. 2010b)

$$\mathbf{\Phi}(\boldsymbol{\lambda}, i) = \varphi(\Psi_0) - \varphi(\Psi_i), \tag{10.17}$$

where $\boldsymbol{\theta} = \boldsymbol{\lambda}$ is a vector of singular values, $\mathbf{\Phi}$ is the primary residual function and $\varphi(\cdot)$ is a function for computing the singular values.

Denoting $\boldsymbol{\theta}_0$ and $\boldsymbol{\theta}$ as parameter vectors that represent normal and abnormal process behaviour, respectively, the statistical LA assumes the following relationship between them (Basseville 1998)

$$\boldsymbol{\theta} = \boldsymbol{\theta}_0 + \frac{\Delta\boldsymbol{\theta}}{\sqrt{m}}, \tag{10.18}$$

where $\Delta\boldsymbol{\theta}$ is a fixed but otherwise unknown vector. The hypotheses that the process behaves normally, H_0 and abnormally, H_1 are as follows

$$\begin{aligned} H_0 &: \boldsymbol{\theta} = \boldsymbol{\theta}_0 \\ H_1 &: \boldsymbol{\theta} = \boldsymbol{\theta}_0 + \frac{\Delta\boldsymbol{\theta}}{\sqrt{m}}, \end{aligned} \tag{10.19}$$

where m is the size of the moving window. The central limit theorem implies that the probability density functions of $\mathbf{\Phi}(\mathbf{\theta}_0,t)$ and $\mathbf{\Phi}(\mathbf{\theta}_0 + \Delta\mathbf{\theta}/\sqrt{m}, i)$ hypothesis asymptotically follow a Gaussian distribution

$$\zeta_0 \triangleq N[\mathbf{0}, \mathbf{\Sigma}(\mathbf{\theta}_0)] \quad \zeta \triangleq N[\mathbf{M}(\mathbf{\theta}_0)\Delta\mathbf{\theta}, \mathbf{\Sigma}(\mathbf{\theta}_0)] \tag{10.20}$$

$$\mathbf{M}(\mathbf{\theta}_0) \triangleq -\frac{\partial\mathbf{\Phi}(\mathbf{\theta}_0, i)}{\partial\mathbf{\theta}}|_{\mathbf{\theta}=\mathbf{\theta}_0}, \tag{10.21}$$

where ζ refers to the improved residual vector, thus

$$\zeta_0 \triangleq \frac{1}{\sqrt{m}} \sum_i^m \mathbf{\Phi}(\mathbf{\theta}_0, i). \tag{10.22}$$

Small departure of the parameter vector $\mathbf{\theta}_0 \rightarrow \mathbf{\theta}$ results in $\zeta_0 \rightarrow \zeta$, and can be identified by a change in mean value. If the process behaves normal, the expectation of the mean value is zero, otherwise it changes to $\mathbf{M}(\mathbf{\theta}_0)\Delta\mathbf{\theta}$. It should be noted, however, that the covariance matrices for $\boldsymbol{\zeta}$ and $\boldsymbol{\zeta}$ are equal and denoted here as $\mathbf{\Sigma}(\mathbf{\theta}_0)$.

Utilizing the improved residual vector $\boldsymbol{\zeta}_0$, allows the construction of a univariate monitoring statistic T^2 (Ge et al. 2010b).

$$T^2(k) = \boldsymbol{\zeta}_0^T(k)\mathbf{\Sigma}^{-1}(\mathbf{\theta}_0)\boldsymbol{\zeta}_0(k). \tag{10.23}$$

This statistic follows a χ^2, which allows a simple calculation of the confidence limit T_α^2. A second monitoring statistic that relates to the Kalman innovation sequences can be established

$$Q(k) = \mathbf{e}^T(k)\mathbf{e}(k) = [\mathbf{z}(k) - \mathbf{C}\mathbf{x}(k)]^T[\mathbf{z}(k) - \mathbf{C}\mathbf{x}(k)]. \tag{10.24}$$

As the Kalman innovation sequences $\mathbf{e}(k)$ follow a Gaussian distribution, the confidence limit, $Q\alpha$ can also be computed easily.

10.6 Application Study in a Gearbox System

This section contrasts the monitoring performance of the two developed monitoring methods. To present a detailed benchmark study with existing work, this comparison also includes conventional principal component analysis (PCA) and ICA methods. From the gearbox system, described in Sect. 10.3, a number of recorded sequences are available, including a reference set describing a fault-free condition and a total of five datasets that represent a gradual removal of one tooth in increments of 10 %. Therefore, the fault conditions describe a 10, 20, 30, 40 and 50 % removal of one tooth. However, the presented comparison here concentrates on the incipient fault conditions 10 and 20 % to study the sensitivity of each method. Each dataset contains 2,620 recorded samples.

The reference set was divided into two equal parts where the first one was used to identify the monitoring models and the other to validate their performance. The

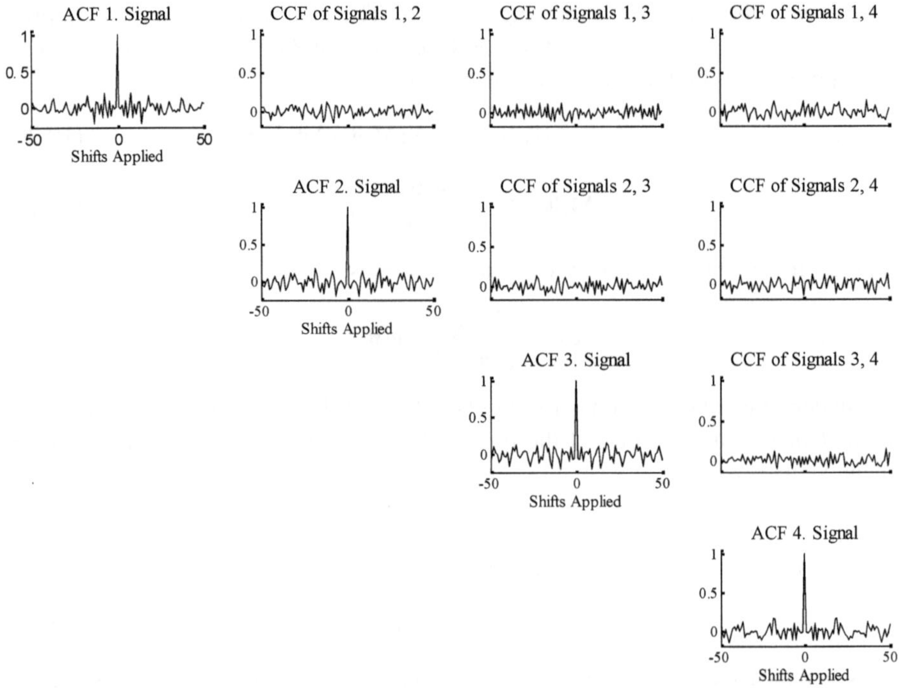

Fig. 10.5 Auto-correlation and cross-correlation functions of the Kalman innovation sequences

monitoring models included the determination of the data representations as well as the statistical confidence limits for each univariate monitoring statistic. For the PCA and ICA model, we used the eigenvector-one rule and the JB test (Jarque and Bera 1987) to determine the number of retained components. Both tests suggested retaining three components for the PCA and ICA models. For DICA, applying the recommendations by Ku et al. (1995), we found that two time-lagged arrangements are sufficient to capture the dynamics in the recorded vibration signals. Again, applying the JB test yielded that the DICA model required the retention of five components. The subsequent analysis of the SVDD approach yielded that the parameters $C=2$ and $\sigma=0.01$ produced a 99 % confidence limit for the transformed ICs. Next, for determining a subspace identification model, the application of the AIC cost function, as proposed by Xie et al. (2006), suggested the selection of 14 state variables. Figure 10.5 points out that the auto- and cross-correlation functions of the residuals of the Kalman innovation model are negligible compared to those of the recorded vibration signals (Fig. 10.3). Finally, we selected three different window length, $m=200$, $m=500$ and $m=800$, for implementing the statistical LA as discussed in the previous section. This is to study the impact of different window lengths upon the sensitivity of the monitoring scheme.

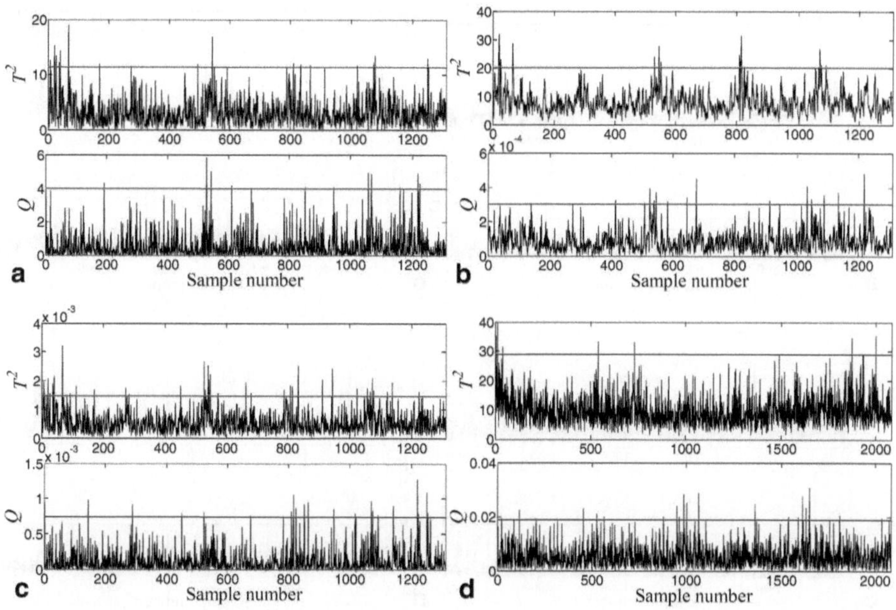

Fig. 10.6 Monitoring results for the reference condition, **a** PCA, **b** ICA, **c** DICA-SVDD, **d** SMILA

Figures 10.6, 10.7 and 10.8 show the monitoring results of the four methods, PCA, ICA, DICA-SVDD and SMILA for the validation data and the 10 and 20 % fault conditions. Although, several data samples exceed the confidence limit, the comparison in Fig. 10.6 highlights that none of the monitoring approaches indicated an out-of-statistical-control situation. This is because we have used the 99 % confidence level, which means 1 % of the data samples are allowed to be out of the normal operation region. However, it is interesting to note that significant trends can be noticed for PCA and ICA. Inspecting Fig. 10.2, particularly sensor 3, showed low frequency trends, which was revealed by a direct analysis of the data using PCA and ICA. Analyzing Figs. 10.7 and 10.8, which describe the performance of each method to the 10 and 20 % fault condition, confirmed that a dynamic monitoring model that can describe stochastic non-Gaussian variables were sensitive to both fault conditions. In contrast, the conventional use of PCA and ICA did not indicate an out-of-statistical-control situation for the 10 % fault. For the 20 % fault condition, PCA produced a statistically significant number of violations for the T^2 statistic, whilst both univariate ICA statistics were not sensitive to this fault. However, the lower plots in Fig. 10.8 showed that the T^2 statistics for the DICA-SVDD and SMILA methods generated a very substantial number of violations with respect to their 99 % confidence limits. Therefore, for both fault conditions, it can be concluded that the developed dynamic methods for non-Gaussian vibration signals are more sensitive in detecting incipient fault conditions using recorded vibration signals. A direct comparison between them suggests that the SMILA technique is more sensitive with regards to this particular

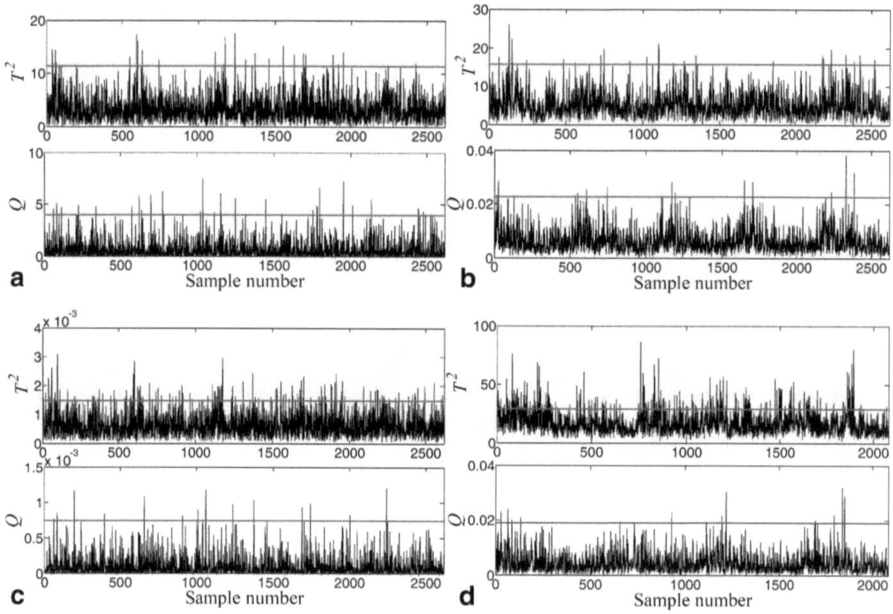

Fig. 10.7 Monitoring results for the faulty condition 10 %, **a** PCA, **b** ICA, **c** DICA-SVDD, **d** SMILA

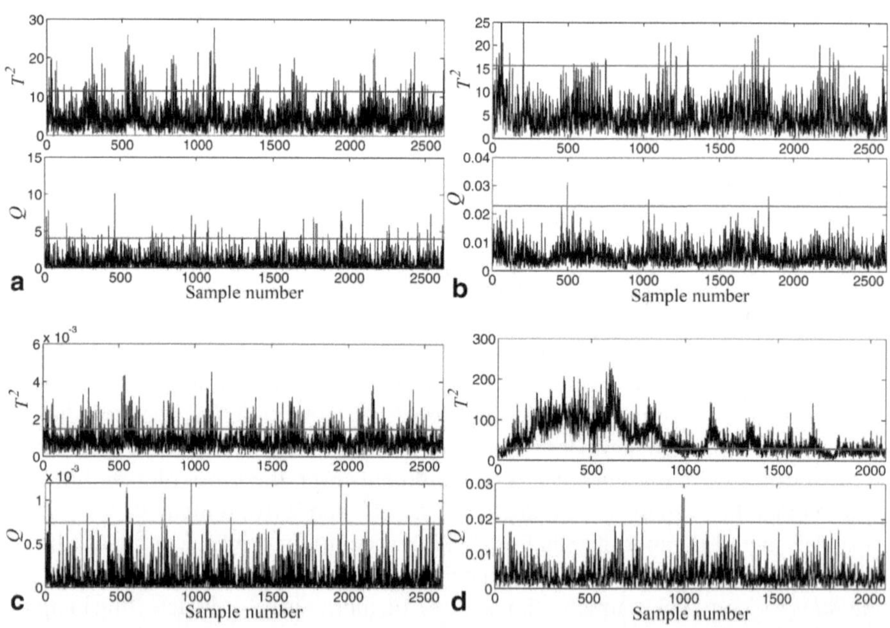

Fig. 10.8 Monitoring results for the faulty condition 20 %, **a** PCA, **b** ICA, **c** DICA-SVDD, **d** SMILA

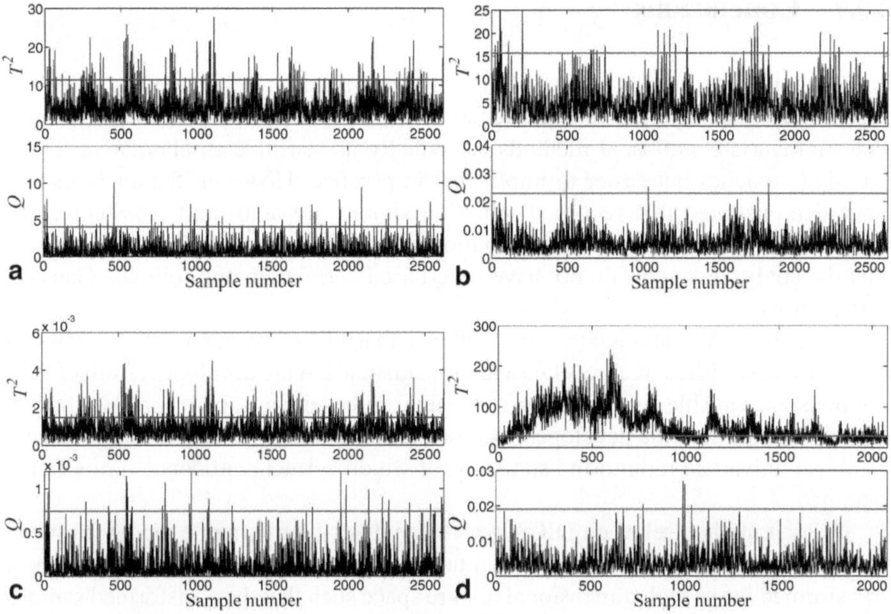

Fig. 10.9 Monitoring results for SMILA approach based on a window size of 800 **a** fault condition 10 %, **b** fault condition 20 %; and 200 **c** fault condition 10 %, **d** fault condition 20 %

application study but a general and definitive statement cannot be made on the basis of this study alone. It is also interesting to note that the Q statistics did not show the same degree of sensitivity as the T^2 ones. More precisely, only the Q statistic for PCA that represents the 20 % fault condition was statistically significant.

The final study here is to examine the influence of the window length upon the performance of the SMILA approach. This is an important issue, as the statistical LA relies on the central limit theorem. Therefore, the larger the window size the more accurately the improved residuals are expected to approximate a multivariate Gaussian distribution. Figure 10.9 shows the results of the SMILA monitoring method for two different window sizes that included 200 and 800 samples. The upper plots refer to a window length of 800 for the 10 % fault condition (left plot) and 20 % fault (right plot), whilst the lower plots show the same conditions for a window length of 200. Whilst both window lengths produced T^2 statistics that were sensitive to each fault condition, the SMILA method based on the larger window length was significantly more sensitive. Comparing Figs. 10.7, 10.8 and 10.9, only a marginal difference could be noticed between a window length of 500 or 800.

10.7 Conclusions

This chapter has introduced two efficient dynamic monitoring methods, by taking a mechanical system as a particular example. The research literature showed that multivariate statistical methods are equally as sensitive as alternative signal-based approaches but easier to implement in practice. However, the analyses here have demonstrated that recorded vibration signals are auto- and cross-correlated and may not follow a Gaussian distribution. Multivariate statistical methods assume that the analyzed signals do not have time-based correlation and follow a Gaussian distribution.

To address this deficiency, two methods that utilize recent work on dynamic multivariate statistics, ICA, SVDD and the statistical LA are developed. Both of these methods are capable of producing a dynamic representation of recorded vibration signals. In addition, the developed methods transform non-Gaussian components to establish univariate monitoring statistics for which correct confidence limits can be obtained.

The first method relies on DICA which can describe a dynamic model for which the source signals are non-Gaussian. Furthermore, these source signals have been transformed into a high dimensional feature space such that the transformed samples fall within a sphere of minimum volume, determined by an SVDD. This method has been abbreviated here as DICA-SVDD. The second method utilizes a Kalman innovation model, which is a dynamic state space representation. The statistical LA has been incorporated into this model to construct a univariate monitoring statistic that follows a parametric chi-squared or F-distribution. The use of SMI allows the determination of the Kalman innovation model and this second approach has been denoted as SMILA.

The performance of both methods has been contrasted with conventional PCA and ICA using recorded data from an experimental test rig that involves a gearbox system. A total of six different conditions, including a reference condition and five fault conditions, have been recorded and analyzed. This benchmark study has shown that the two developed methods could detect incipient fault conditions, where 10 % and 20 % of the length of one tooth has been removed, whilst PCA and ICA failed to detect the 10 % condition. Moreover, the direct comparison between both the developed methods has suggested that the SMILA method is more sensitive although a general conclusion cannot be drawn from a single application study.

Chapter 11
Probabilistic Process Monitoring

11.1 Introduction

In modern industrial processes, a huge number of process data have been collected by the widely used distributed control system (DCS), which makes the data-based method very popular for process modeling and monitoring. As a represented data-based method for process monitoring, statistical-based approaches such as principal component analysis (PCA) and partial least squares (PLS) have caught much attention since the last decade (Qin 2003; Wang et al. 2003; Kruger and Dimitriadis 2008). Extensions of the traditional statistical-based method have also been developed, such as nonlinear PCA, multiscale PCA, dynamic PCA, etc. (Zhang and Qin 2008; Bakshi 1998; Maulud et al. 2006; Ku et al. 1995; Yao and Gao 2009; AlGhazzawi and Lennox 2008).

In practical applications, process data (normal or not) are always contaminated by noises. To address this problem, a probabilistic counterpart of PCA (probabilistic principal component analysis, [PPCA]) has recently been developed for process monitoring (Kim and Lee 2003). PPCA was originally proposed by Tipping and Bishop (1999). It can not only present the model in a probabilistic manner upon which statistical judgments and decisions can be made, but can also catch the noise information simultaneously.

An important issue of the PPCA method is how to determine the retained number of principal components. In fact, the selection of the latent variable dimensionality can be termed as a model complexity problem, which can be addressed by the Bayesian method. Particularly, a Bayesian regularization of PCA (Bayesian PCA [BPCA]) has been proposed by Bishop (1999), and utilized in many areas, such as image processing, spectroscopic data analysis, etc. Later, a more general modeling framework of the BPCA algorithm was proposed, in which different aspects among the Bayesian model structure have been detailed, such as the prior density function, the loss function, the model rank, etc. (Nounou et al. 2002). However, to our best knowledge, it has not been used for process monitoring. Here, we intend to introduce the Bayesian PCA method into the process monitoring area, thus a BPCA-based monitoring approach is constructed. Different from PPCA, the BPCA

Z. Ge, Z. Song, *Multivariate Statistical Process Control,* Advances in Industrial Control,
DOI 10.1007/978-1-4471-4513-4_11,

can automatically determine the dimensionality of the latent variable space when constructing the probabilistic model. Precisely, the dimensionality of the retained principal components is controlled by a hyperparameter in the BPCA model, which also has its probabilistic interpretation for each principal component. Besides, based on the Bayesian treatment, the new monitoring approach has more advantages when the number of modeling data samples is limited. This is because it can avoid singularity and overfit problems by suppressing unwanted degrees of freedom in the model structure, both of which could be caused by the traditional PPCA and PCA methods.

For monitoring processes with multiple operation conditions under the noisy environment, the PPCA-based method has been extended to its mixture form (Tipping and Bishop 1999; Chen and Liu 1999, 2000; Choi et al. 2004, 2005; Yu and Qin 2008, 2009; Chen and Zhang 2010; Ge and Song 2010a). However, there are several drawbacks of the (mixture) PPCA-based methods. First, they cannot determine the effective principal component number automatically through the modeling process. Second, the importance of each retained principal component in the model structure cannot be easily differentiated. Besides, singularity and overfit problems may also be caused when the number of modeling data samples is limited.

Based on the BPCA model, a mixture form can be developed, thus a mixture Bayesian regularization model of PPCA (MBPCA) is constructed here (Ge and Song 2010b). Then, a corresponding multimode monitoring scheme can be formulated. For monitoring a new process data sample, the traditional mixture PPCA method first calculates its posterior probability in different operation modes, then the submodel with the largest posterior probability value is selected for monitoring. However, when two or several posterior probability values are comparative, the model selection will be confused, which may decrease the monitoring performance. To address this issue, a probabilistic combination strategy has been proposed to enhance the multimode monitoring performance (Ge and Song 2010b). Precisely, a weighted monitoring scheme was developed, that is, the operation mode with a higher posterior probability value will be given a larger weight through the combination step. Therefore, the confused model selection problem is avoided. Instead, all monitoring results in different operation modes have been utilized to make the final decision, which is more reliable. Besides, another important aspect is the mode localization problem. That is, one may want to know the mode information of the monitored data sample, i.e., which operation mode it belongs to. Depending on the MBPCA-based monitoring approach, a mode localization approach has also been developed, which can successfully locate the monitored data sample to its correct operation mode.

11.2 PPCA-Based Method for Process Monitoring

As a probabilistic counterpart of PCA, the PPCA was first proposed by Tipping and Bishop (1999), and later adopted for process monitoring purpose (Kim and Lee 2003). Through a generative model structure, the formulation of PPCA can be given as

$$\mathbf{x} = \mathbf{Pt} + \mathbf{e} \tag{11.1}$$

where $\mathbf{x} \in R^m$ represents the process variable, $\mathrm{t} \in R^k$ is the latent variable, $\mathbf{P} \in R^{m \times k}$ is the loading matrix, and $\mathbf{e} \in R^m$ is a zero mean white noise term with variance $\beta^{-1}\mathbf{I}$, thus $p(\mathbf{e}) = N(\mathbf{e}|0, \beta^{-1}\mathbf{I})$. In the PPCA model, the prior distribution of the latent variable $\mathbf{t}$ is also assumed to be a Gaussian distribution with zero mean and one variance $p(\mathbf{t}) = N(\mathbf{t}|0, \mathbf{I})$. Therefore, the marginal likelihood of $\mathbf{x}$ can be calculated as

$$p(\mathbf{x}|\mathbf{P}, \beta) = \int p(\mathbf{x}|\mathbf{t}, \mathbf{P}, \beta) p(\mathbf{t}) d\mathbf{t} \tag{11.2}$$

For a given dataset $\mathrm{X} = (\mathrm{x}_1, \mathrm{x}_2, \ldots, \mathrm{x}_n)$ of n data samples, $\mathbf{P}$ and β can be determined by maximizing the following likelihood function through an Expectation-Maximization (EM) algorithm

$$L(\mathbf{P}, \beta) = \ln \prod_{i=1}^{n} p(\mathbf{x}_i|\mathbf{P}, \beta) \tag{11.3}$$

After the parameter set of PPCA has been determined, the corresponding monitoring scheme can be developed. For a new process data sample x_{new}, the latent variable of this sample can be calculated as

$$\mathbf{t}_{new} = \mathbf{Q}\mathbf{x}_{new} = \mathbf{P}^T (\mathbf{P}\mathbf{P}^T + \Sigma)^{-1} \mathbf{x}_{new} \tag{11.4}$$

The estimated variance of the latent variable is given in Eq. (11.5), through which we can find that it has no relationship with the current data sample x_{new}

$$\mathrm{var}\,(\mathbf{t}_{new}) = \mathbf{Q}(\mathbf{P}\mathbf{P}^T + \Sigma)\mathbf{Q}^T \tag{11.5}$$

Therefore, the T^2 statistic can be constructed as

$$T^2_{new} = \mathbf{t}^T_{new}(\mathrm{var}\,(\mathbf{t}_{new}))^{-1}\mathbf{t}_{new} \tag{11.6}$$

Similarly, the *SPE* statistic can be constructed as

$$SPE_{new} = \mathbf{e}^T_{new}(\beta^{-1}\mathbf{I})^{-1}\mathbf{e}_{new} \tag{11.7}$$

where $\mathbf{e}_{new} = \mathbf{x}_{new} - \mathbf{P}\mathbf{t}_{new} = (\mathbf{I} - \mathbf{PQ})\mathbf{x}_{new}$ is the error term of the new data sample. The control limits of both monitoring statistics can be determined by the χ^2 distribution with appropriate dimensions of freedom, thus $T^2_{\lim} = \chi^2_\gamma(k)$, $SPE_{\lim} = \chi^2_\gamma(m)$, where γ is the significant level (Kim and Lee 2003).

It is noted that an important assumption of PPCA is that the dimension of the latent variable k is known beforehand. In fact, if the number of process data samples is limited, the selection of principal component number will become problematic. This is because the PPCA method itself does not provide any mechanism to determine the effective latent variable dimensionality. If there are not enough data samples available for cross-validation, it is difficult to determine this important number. Therefore, it is desired that the number of effective latent variables could be determined automatically through the model development step, especially when training data samples are limited. Fortunately, this problem can be well solved by the Bayesian regularization method, which is detailed in the next section.

11.3 Bayesian Regularization of PPCA Monitoring Method

Depending on the probabilistic formulation of PPCA defined in Eq. (11.1), the Bayesian treatment of PCA is obtained by introducing a prior distribution over the parameter set $\{\mathbf{P},\beta\}$, represented as $p(\mathbf{P},\beta)$. Then, the posterior distribution $p(\mathbf{P},\beta|\mathbf{X})$ can be obtained by multiplying the prior by the likelihood function, whose logarithm is given in Eq. (11.3). To implement the Bayesian PCA algorithm, two important issues should be addressed. The first one is the choice of prior distribution for the parameter set, and the second one is how to develop a tractable algorithm to determine the final model structure.

First, the dimensionality of the latent space is set to its maximum value $d=m-1$ Then, a hyperparameter vector $\alpha=\{\alpha_1,\alpha_2,\ldots,\alpha_d\}$ is introduced to control the dimensionality of the latent space. Though a conditional Gaussian distribution, each hyperparameter controls one column of the loading matrix $\mathbf{P}$, which is defined as (Bishop 1999, 2006)

$$p(\mathbf{P}|\alpha)=\prod_{i=1}^{d}\left(\frac{\alpha_i}{2\pi}\right)^{m/2}\exp\left\{-\frac{1}{2}\alpha_i\|\mathbf{p}_i\|^2\right\} \tag{11.8}$$

where $\mathbf{p}_i$is the ith column of the loading matrix $\mathbf{P}$. As each α_i controls the inverse variance of $\mathbf{p}_i$, if it has a large value, the corresponding $\mathbf{p}_i$ will tend to be very small, and would be effectively removed from the latent space loading matrix.

For the purpose of controlling the effective dimensionality of the latent space, it is sufficient to estimate an optimal value of $\mathbf{P}$ or β, which can be determined by using the EM algorithm. Thus, by maximizing $\mathbf{P}$ or β in the posterior distribution function given in Eq. (11.9), the EM algorithm can be calculated as follows (Bishop 1999)

$$\ln p(\mathbf{P}|\mathbf{X})=L-\frac{1}{2}\sum_{i=1}^{m-1}\alpha_i\|\mathbf{p}_i\|^2+const \tag{11.9}$$

In the E-step, the expected sufficient statistics of the latent variable can be evaluated as

$$E(\hat{\mathbf{t}}|\mathbf{x})=E(\hat{\mathbf{t}})+\mathrm{cov}\,(\hat{\mathbf{t}},\mathbf{x})[\,\mathrm{var}\,(\mathbf{x})]^{-1}[\mathbf{x}-E(\mathbf{x})] \tag{11.10}$$

$$\begin{aligned} E(\hat{\mathbf{t}}\hat{\mathbf{t}}^T|\mathbf{x})&=\mathrm{var}\,(\hat{\mathbf{t}}|\mathbf{x})+E(\hat{\mathbf{t}}|\mathbf{x})E^T(\hat{\mathbf{t}}|\mathbf{x})\\ \mathrm{var}\,(\hat{\mathbf{t}}|\mathbf{x})&=\mathrm{var}\,(\hat{\mathbf{t}})-\mathrm{cov}\,(\hat{\mathbf{t}},\mathbf{x})\mathrm{var}^{-1}(\mathbf{x})\,\mathrm{cov}\,(\mathbf{x},\hat{\mathbf{t}}) \end{aligned} \tag{11.11}$$

where $E(\hat{\mathbf{t}}) = 0$, var $(\mathbf{x}) = \mathbf{P}\mathbf{P}^T + \beta^{-1}\mathbf{I}$, $E(\mathbf{x}) = 0$, var $(\hat{\mathbf{t}}) = \mathbf{I}$, and

$$\text{cov}\,(\hat{\mathbf{t}}, \mathbf{x}) = E(\hat{\mathbf{t}}\mathbf{x}^T) - E(\hat{\mathbf{t}})E(\mathbf{x}^T) = \mathbf{P}^T \quad (11.12)$$

$$\text{cov}\,(\mathbf{x}, \hat{\mathbf{t}}) = E(\mathbf{x}\hat{\mathbf{t}}^T) - E(\mathbf{x})E(\hat{\mathbf{t}}^T) = \mathbf{P} \quad (11.13)$$

Then, Eqs. (11.10) and (11.11) become as follows

$$\begin{aligned} E(\hat{\mathbf{t}}|\mathbf{x}) &= \mathbf{M}^{-1}\mathbf{P}^T\mathbf{x} \\ E(\hat{\mathbf{t}}\hat{\mathbf{t}}^T|\mathbf{x}) &= \beta^{-1}\mathbf{M}^{-1} + E(\hat{\mathbf{t}}|\mathbf{x})E^T(\hat{\mathbf{t}}|\mathbf{x}) \end{aligned} \quad (11.14)$$

where $\mathbf{M} = \mathbf{P}^T\mathbf{P} + \beta^{-1}\mathbf{I}$. The M-step then updates the model parameters by maximizing the log posterior distribution function with respect to $\mathbf{P}$ and β, and setting them to zero, thus

$$\frac{\partial[\ln p(\mathbf{P}|\mathbf{X})]}{\partial \mathbf{P}} = 0 \quad (11.15)$$

$$\frac{\partial[\ln p(\mathbf{P}|\mathbf{X})]}{\partial \beta} = 0 \quad (11.16)$$

Then, the updated parameters can be given as follows

$$\begin{aligned} \hat{\mathbf{P}} &= \left[\sum_{i=1}^{n} E(\hat{\mathbf{t}}_i|\mathbf{x}_i)\mathbf{x}_i^T\right]\left[\sum_{i=1}^{n} E(\hat{\mathbf{t}}_i\hat{\mathbf{t}}_i^T|\mathbf{x}_i) + \beta^{-1}\mathbf{A}\right]^{-1} \\ \hat{\alpha}_i &= \frac{m}{\|\mathbf{p}_i\|^2} \\ \hat{\beta} &= \frac{nm}{\sum_{i=1}^{n}\{\mathbf{x}_i^T\mathbf{x}_i - 2E^T(\hat{\mathbf{t}}_i|\mathbf{x}_i)\hat{\mathbf{P}}^T\mathbf{x}_i + Tr[E(\hat{\mathbf{t}}_i\hat{\mathbf{t}}_i^T|\mathbf{x}_i)\hat{\mathbf{P}}^T\hat{\mathbf{P}}]\}} \end{aligned} \quad (11.17)$$

where $\mathbf{A} = diag(\alpha_i, i = 1, 2, \ldots, d)$ is a diagonal matrix, $Tr(\cdot)$ is an operator for trace value calculation. Therefore, the optimal values of $\mathbf{P}$ and β can be determined by recursively calculating E-step and M-step given above, until all the parameters satisfy a suitable convergence criterion, such as 10^{-4}.

After the BPCA model structure has been constructed, the monitoring model can be developed. Suppose a new data sample $\mathbf{x}_{new}$ has been collected, its latent variable vector can be estimated as

$$\hat{\mathbf{t}}_{new} = \mathbf{P}^T(\mathbf{P}\mathbf{P}^T + \beta^{-1}\mathbf{I})^{-1}\mathbf{x}_{new} \quad (11.18)$$

Then, the mean prediction error can be calculated as

$$\hat{\mathbf{e}}_{new} = \mathbf{x}_{new} - \mathbf{P}\hat{\mathbf{t}}_{new} = [\mathbf{I} - \mathbf{P}\mathbf{P}^T(\mathbf{P}\mathbf{P}^T + \beta^{-1}\mathbf{I})^{-1}]\mathbf{x}_{new} \quad (11.19)$$

To monitor the latent space and the modeling error (noisy space) separately, two statistics can be constructed as follows

$$T_{new}^2 = \hat{\mathbf{t}}_{new}^T \mathrm{var}^{-1}(\mathbf{t}_{new}|\mathbf{x}_{new})\hat{\mathbf{t}}_{new} \tag{11.20}$$

$$SPE_{new} = \hat{\mathbf{e}}_{new}^T \hat{\mathbf{e}}_{new} \tag{11.21}$$

where $\mathrm{var}(\mathbf{t}_{new}|\mathbf{x}_{new}) = \beta^{-1}\mathbf{M}^{-1}$ is the variance of the latent variable. The control limits of the T^2 and SPE statistics can both be determined by the χ^2 distribution with different degrees of freedom, thus (Zhang et al. 2010)

$$T_{new}^2 \leq T_{\lim}^2 = \chi_\gamma^2(k) \tag{11.22}$$

$$\begin{aligned} SPE_{new} &\leq SPE_{\lim} = g \cdot \chi_{\gamma,h}^2 \\ g \cdot h &= mean(SPE_{tr}) \\ 2g^2 h &= \mathrm{var}\,(SPE_{tr}) \end{aligned} \tag{11.23}$$

where SPE_{tr} represents the SPE value set of the train dataset, $mean(SPE_{tr})$ is the mean value of SPE_{tr}, $\mathrm{var}(SPE_{tr})$ is the variance of SPE_{tr}, k is the selected number of retained principal components in the Bayesian PCA model, and γ is the significance level of both monitoring statistics.

11.4 Multimode Process Monitoring Based on MBPCA

11.4.1 Mixture Bayesian Regularization of PPCA

When the process dataset comes from several different operation modes, the single Bayesian PCA-based modeling method is not sufficient. Hence, this section extends the Bayesian regularization treatment of PPCA to its mixture form, thus a mixture Bayesian regularization method of PPCA is developed. Suppose a total of C operation conditions are incorporated, the distribution of the process variable can be calculated as

$$p(\mathbf{x}) = \sum_{c=1}^{C} p(\mathbf{x}|c)p(c) \tag{11.24}$$

where $p(c)$ is the mixing proportion of each operation mode, under the constraint that $\sum_{c=1}^{C} p(c) = 1$, $p(\mathbf{x}|c)$ is the conditional distribution function of each operation mode, which follows Gaussian distribution as $N(0, \mathbf{P}_c\mathbf{P}_c^T + \beta_c^{-1}\mathbf{I})$, where $\mathbf{P}_c$ and β_c are the corresponding parameters in the cth local Bayesian PCA model. For Bayesian regularization, a hyperparameter matrix is defined as follows (Ge and Song 2010b)

$$\alpha = \begin{bmatrix} \alpha_{11} & \alpha_{12} & \cdots & \alpha_{1d} \\ \alpha_{21} & \alpha_{22} & \cdots & \alpha_{2d} \\ \vdots & \vdots & \alpha_{ci} & \vdots \\ \alpha_{C1} & \alpha_{C2} & \cdots & \alpha_{Cd} \end{bmatrix} \tag{11.25}$$

where $c=1,2,\ldots,C$, $i=1,2,\ldots,d=m-1$. Then, the aim of the mixture Bayesian PCA model development can be formulated as follows: given a process dataset $\mathbf{X} \in R^{n\times m}$ estimate the parameter sets $\mathbf{\Theta_c} = \{\mathbf{P}_c,\beta_c,\alpha_c\}(c=1,2,\ldots,C)$ for each local Bayesian PCA model.

First, the log posterior distribution function is given as

$$\begin{aligned} \ln p(\mathbf{P}|\mathbf{X}) &= \ln \prod_{j=1}^{n} p(\mathbf{x}_j|\mathbf{P}) - \frac{1}{2}\sum_{i=1}^{d} \alpha_{ci}\, p(c)\|\mathbf{p}_i\|^2 + const \\ &= \sum_{j=1}^{n} \ln\left[\sum_{c=1}^{C} p(\mathbf{x}_j|c,\mathbf{P})p(c)\right] - \frac{1}{2}\sum_{i=1}^{d} \alpha_{ci}\, p(c)\|\mathbf{p}_i\|^2 + const \end{aligned} \tag{11.26}$$

Then, a new EM algorithm can be constructed for estimation of the parameter set $\mathbf{\Theta_c}$ in the mixture Bayesian PCA model. In the E-step, we are given the parameters $\mathbf{\Theta}_{old}$, obtained in the previous step, our aim is to estimate the expected sufficient statistics of the latent variable posterior distribution and the mode posterior probability $p(c|\mathbf{x},\mathbf{\Theta}_{old})$. Through the Bayesian rule, the posterior probabilities of the mode and the latent variable can be calculated as follows

$$p(c|\mathbf{x},\Theta_{old}) = \frac{p(\mathbf{x}|c,\Theta_{old})p(c|\Theta_{old})}{p(\mathbf{x}|\Theta_{old})} \tag{11.27}$$

$$p(\mathbf{t}|\mathbf{x},c,\Theta_{old}) = \frac{p(\mathbf{x}|\mathbf{t},c,\Theta_{old})p(\mathbf{t}|c,\Theta_{old})}{p(\mathbf{x}|c,\Theta_{old})} \tag{11.28}$$

where $p(c|\mathbf{\Theta}_{old})$ is the proportion value calculated in the previous step and $p(\mathbf{x}_i|c,\mathbf{\Theta}_{old})$ follows the multivariate Gaussian distribution $N(0,\mathbf{P}_c\mathbf{P}_c^T + \beta_c^{-1}\mathbf{I})$. The estimated mean and variance values of the latent variable are given as

$$\begin{aligned} E(\hat{\mathbf{t}}|\mathbf{x},c,\Theta_{old}) &= \mathbf{M}_c^{-1}\mathbf{P}_c^T\mathbf{x} \\ E(\hat{\mathbf{t}}\hat{\mathbf{t}}^T|\mathbf{x},c,\Theta_{old}) &= \beta_c^{-1}\mathbf{M}_c^{-1} + E(\hat{\mathbf{t}}|\mathbf{x},c,\Theta_{old})E^T(\hat{\mathbf{t}}|\mathbf{x},c,\Theta_{old}) \end{aligned} \tag{11.29}$$

where $\mathbf{M}_c = \mathbf{P}_c^T\mathbf{P}_c + \beta_c^{-1}\mathbf{I}$. The M-step involves estimating the optimal parameter values through the maximum likelihood method, which are given as (Ge and Song 2010b)

$$p(c|\Theta_{new}) = \frac{1}{n}\sum_{j=1}^{n} p(c|\mathbf{x}_j,\Theta_{old}) \tag{11.30}$$

$$\mathbf{P}_c^{new} = \left(\sum_{j=1}^{n} p(c|\mathbf{x}_j, \Theta_{old}) \mathbf{x}_j E(\mathbf{t}_j^T | \mathbf{x}_j, c, \Theta_{old}) \right) \times \left(\sum_{j=1}^{n} p(c|\mathbf{x}_j, \Theta_{old}) E(\mathbf{t}_j \mathbf{t}_j^T | \mathbf{x}_j, c, \Theta_{old}) + p(c|\Theta_{new}) \beta_c^{-1} \mathbf{A}_c \right)^{-1} \quad (11.31)$$

$$\alpha_{ci}^{new} = \frac{m}{\|\mathbf{p}_{ci}^{new}\|^2} \quad (11.32)$$

$$\beta_c^{new} = \frac{nm}{\sum_{j=1}^{n} p(c|\mathbf{x}_j, \Theta_{old}) \{ \mathbf{x}_j^T \mathbf{x}_j - 2E^T(\mathbf{t}_j|\mathbf{x}_j) \mathbf{P}_c^{newT} \mathbf{x}_j + Tr[E(\mathbf{t}_j \mathbf{t}_j^T | \mathbf{x}_j) \mathbf{P}_c^{newT} \mathbf{P}_c^{new}] \}} \quad (11.33)$$

where $\mathrm{A_c} = diag(\alpha_{ci}, i = 1,2,\ldots,d)$ is a diagonal matrix, $c = 1,2,\ldots,C$, $\mathbf{p}_{ci}^{new}$ is the i-th column of the new estimated loading matrix $\mathbf{P}_c^{new}$. Detailed derivation of the EM algorithm for mixture BPCA model is provided in Appendix. Updating the E-step and the M-step recursively, we can finally get an optimal value of the parameter set $\Theta_c^{opt} = \{\hat{\mathbf{P}}_c, \hat{\beta}_c, \hat{\alpha}_c\} (c = 1, 2, \cdots, C)$ for each local Bayesian PCA model. Thus, the mixture Bayesian regularization model of PPCA has been constructed. Different from the traditional mixture PPCA model, the effective number of principal components in each local BPCA model plane can be determined automatically by the new method.

11.4.2 Multimode Process Monitoring

Based upon the mixture BPCA model in Sect. 11.4.1, a specific monitoring scheme for multimode processes can be constructed. Suppose a new data sample $\mathbf{x}_{new}$ has been collected from the process, we first construct two traditional monitoring statistic (T^2 and SPE) in each of the C local BPCA model spaces, which can be calculated as follows

$$\mathbf{t}_{c,new} = \hat{\mathbf{Q}}_c \mathbf{x}_{new}$$
$$T^2_{c,new} = \mathbf{t}^T_{c,new} \left(\hat{\mathbf{Q}}_c (\hat{\mathbf{P}}_c \hat{\mathbf{P}}_c^T + \hat{\beta}_c^{-1} \mathbf{I}) \hat{\mathbf{Q}}_c^T \right)^{-1} \mathbf{t}_{c,new} \quad (11.34)$$

$$\mathbf{e}_{c,new} = (\mathbf{I} - \hat{\mathbf{P}}_c \hat{\mathbf{Q}}_c)(\mathbf{x}_{new} - \mu_k)$$
$$SPE_{c,new} = \mathbf{e}^T_{c,new} (\hat{\beta}_c^{-1} \mathbf{I})^{-1} \mathbf{e}_{c,new} \quad (11.35)$$

where $c = 1,2,\ldots,C$, $\hat{\mathbf{Q}}_c = \hat{\mathbf{P}}_c^T (\hat{\mathbf{P}}_c \hat{\mathbf{P}}_c^T + \hat{\beta}_c^{-1} \mathbf{I})^{-1}$. In the mixture PPCA monitoring method, a total of C monitoring charts can be built, the one with the largest posterior probability value is selected for monitoring. Through the Bayesian rule, the posterior probability of each operation mode can be calculated as

$$p(c|\mathbf{x}_{new}, \Theta^{opt}) = \frac{p(\mathbf{x}_{new}|c, \Theta^{opt})p(c|\Theta^{opt})}{p(\mathbf{x}_{new}|\Theta^{opt})} \tag{11.36}$$

Hence, the final T^2 and SPE monitoring statistics of the mixture PPCA method can be determined as

$$T^2_{final,new} = T^2_{c,new}\{\max(\mathrm{p}(\mathrm{c}|\mathbf{x}_{\mathrm{new}}, \Theta^{\mathrm{opt}})), \mathrm{c} = 1, 2, \cdots, \mathrm{C}\} \tag{11.37}$$

$$SPE_{final,new} = SPE_{c,new}\{\max(\mathrm{p}(\mathrm{c}|\mathbf{x}_{\mathrm{new}}, \Theta^{\mathrm{opt}})), \mathrm{c} = 1, 2, \cdots, \mathrm{C}\} \tag{11.38}$$

Then, the process can be monitored by two normalized statistics, the control limits of which are selected as 1 (Choi et al. 2005). However, in our opinion, some useful information may be lost through the hard assignment of the mixture PPCA monitoring method. Besides, if two or more posterior probability values are comparative, it is difficult to make a final monitoring decision. In such a special case, it can be inferred that the monitoring performance may be deteriorated. In contrast, a soft assignment strategy is proposed to enhance the monitoring performance, which is based on Bayesian transformation and probabilistic combination.

As different local BPCA model spaces have their own monitoring statistics and control limits, it is difficult to combine them directly. Here, the Bayesian method is employed to turn the monitoring results into fault probabilities, which are given as

$$P^c_{T^2}(F|\mathbf{x}_{new}) = \frac{P^c_{T^2}(\mathbf{x}_{new}|F)P^c_{T^2}(F)}{P^c_{T^2}(\mathbf{x}_{new})}$$

$$P^c_{T^2}(\mathbf{x}_{new}) = P^c_{T^2}(\mathbf{x}_{new}|N)P^c_{T^2}(N) + P^c_{T^2}(\mathbf{x}_{new}|F)P^c_{T^2}(F) \tag{11.39}$$

$$P^c_{SPE}(F|\mathbf{x}_{new}) = \frac{P^c_{SPE}(\mathbf{x}_{new}|F)P^c_{SPE}(F)}{P^c_{SPE}(\mathbf{x}_{new})}$$

$$P^c_{SPE}(\mathbf{x}_{new}) = P^c_{SPE}(\mathbf{x}_{new}|N)P^c_{SPE}(N) + P^c_{SPE}(\mathbf{x}_{new}|F)P^c_{SPE}(F) \tag{11.40}$$

where $P^c_{T^2}(N)$, $P^c_{T^2}(F)$ and $P^c_{SPE}(N)$, $P^c_{SPE}(F)$ represent the prior probabilities of the normal and faulty process conditions, which can be defined as

$$P^c_{T^2}(N) = P^c_{SPE}(N) = 1 - \gamma$$

$$P^c_{T^2}(F) = P^c_{SPE}(F) = \gamma \tag{11.41}$$

where γ is the significant level. In order to obtain the fault probabilities of the new data sample, we should also know the conditional probabilities, which are defined as follows (Ge and Song 2010b)

$$P^c_{T^2}(\mathbf{x}_{new}|N) = \exp\left\{-\frac{T^2_{c,new}}{T^2_{c,\lim}}\right\} \quad P^c_{T^2}(\mathbf{x}_{new}|F) = \exp\left\{-\frac{T^2_{c,\lim}}{T^2_{c,new}}\right\} \tag{11.42}$$

$$P_{SPE}^c(\mathbf{x}_{new}|N) = \exp\left\{-\frac{SPE_{c,new}}{SPE_{c,\lim}}\right\} \quad P_{SPE}^c(\mathbf{x}_{new}|F) = \exp\left\{-\frac{SPE_{c,\lim}}{SPE_{c,new}}\right\} \tag{11.43}$$

After the monitoring results of the new data sample in each local BPCA model space has been obtained, we can easily calculate the final T^2 and SPE monitoring statistics by combining them through their posterior probabilities, which are given as follows

$$T_{final,new}^2 = \sum_{c=1}^{C} p(c|\mathbf{x}_{new}, \Theta^{opt})P_{T^2}^c(F|\mathbf{x}_{new}) \tag{11.44}$$

$$SPE_{final,new} = \sum_{c=1}^{C} p(c|\mathbf{x}_{new}, \Theta^{opt})P_{SPE}^c(F|\mathbf{x}_{new}) \tag{11.45}$$

Based on these two final monitoring statistics, we can judge the process behavior by examining if these two statistics have violated their corresponding control limits γ. If their values exceed γ, some fault could be detected, otherwise, the process is normal, and the monitoring procedures should be kept on.

As a part of the process monitoring task, mode localization is also important. Knowing the exact operation mode of the monitored data sample, the engineer can easily locate the production of the process, and the problem will also be found if there is any. Here, a mode localization method is developed based on the mixture BPCA monitoring framework. Having calculated the posterior probabilities of the monitored data sample corresponding to different operation modes, we can simply locate the mode information through these posterior probabilities, thus the mode with the largest posterior probability value should be determined as the current operation mode. However, this method may not function well when an unknown operation mode happens in the process. Based on the posterior probability localization method, the data sample from the new operation mode will be assigned to one of the known operation mode, which is not the real case. In other words, mode localization based on the posterior probability is not reliable.

In order to emphasize the reliability for mode localization, the joint probability analysis method has been introduced (Ge and Song 2010b). Similar to the posterior probability, successful localization will be obtained if the data sample was collected under known operation modes. However, different from the posterior probability method, the joint probability analysis can also successfully identify the change of the process. That is, if the process is changed to a new operation mode, the joint probabilities of the monitored data sample with all known operation modes will approach to zero. To get the joint probability values of the new data sample $\mathbf{x}_{new}$ with different operation modes, the conditional probabilities should be calculated first, which follows a multivariate Gaussian distribution. The conditional probability function is given as

$$p(\mathbf{x}|c) = \int p(\mathbf{x}|\mathbf{t}, c)p(\mathbf{t}|c)p(c)d\mathbf{t} \tag{11.46}$$

For the new data sample $\mathbf{x}_{new}$, its conditional probability can be calculated as $p(\mathbf{x}_{new}|c) = p(\mathbf{x}|c)|_{\mathbf{x}_{new}}$. Then, the joint probabilities values of $\mathbf{x}_{new}$ can be given as follows

$$JP_{c,new} = p(\mathbf{x}_{new}|c)p(c|\mathbf{x}_{new}, \Theta^{opt}) \tag{11.47}$$

where $c = 1,2,\ldots,C$. Through analysis of Eq. (11.47), it can be inferred that if the process is changed to a new operation mode, the value of the conditional probability of $\mathbf{x}_{new}$ ($p(\mathbf{x}_{new}|c)$) will be very small. Although the mode posterior probability value $p(c|\mathbf{x}_{new}, \Theta^{opt})$ will not decrease to zero, since they are restricted to $\sum_{c=1}^{C} p(c|\mathbf{x}_{new}, \Theta^{opt}) = 1$, by multiplying the mode posterior probability value with the conditional probability value, the joint probability value of $\mathbf{x}_{new}$ will quickly decrease to zero. Hence, if all joint probability values approach to zero, a new operation mode can be judged.

11.5 Case Studies

In this section, two case studies are provided to evaluate the efficiency of the probabilistic methods. The first one is a numerical example with six variables, and the other one is a real industrial polypropylene production process application case.

11.5.1 A Numerical Example

This numerical example consists of six variables, which are driven by three Gaussian latent variables, the relationships between them are given by

$$\begin{aligned}
\mathbf{x}_1 &= \mathbf{P}_1\mathbf{t}_1 + \mathbf{e}_1, \quad \mathbf{t}_1 \in N(-1, 1), \quad \mathbf{e}_1 \in N(0, 0.1) \\
\mathbf{x}_2 &= \mathbf{P}_2\mathbf{t}_2 + \mathbf{e}_2, \quad \mathbf{t}_2 \in N(0, 1), \quad \mathbf{e}_2 \in N(0, 0.1) \\
\mathbf{x}_3 &= \mathbf{P}_3\mathbf{t}_3 + \mathbf{e}_3, \quad \mathbf{t}_3 \in N(1, 1), \quad \mathbf{e}_3 \in N(0, 0.1)
\end{aligned} \tag{11.48}$$

where $\mathbf{x}_1,\mathbf{x}_2,\mathbf{x}_3 \in R^6$ and $\mathbf{t}_1,\mathbf{t}_2,\mathbf{t}_3 \in R^3$ are Gaussian distributed latent variables with different mean values, and $\mathbf{e}_1,\mathbf{e}_2,\mathbf{e}_3 \in R^6$ are Gaussian process noises with an equal variance 0.1. The three loading matrices $\mathbf{P}_1$, $\mathbf{P}_2$, and $\mathbf{P}_3$ are randomly selected as 6×3 matrices. Therefore, three operation modes can be simulated through Eq. (11.48). To build the mixture BPCA and the mixture PPCA models, 200 data samples of each cluster have been generated, thus a total of 600 data samples are used for model construction. To simulate the process fault, two cases are assumed as follows

Case 1: A step bias by adding 0.5 to the second variable of the first operation mode is introduced starting from sample 101 to 200.

Case 2: A ramp change by adding $0.01(i - 100)$ to the first variable of the second operation mode is introduced starting from sample number $i = 101$ to $i = 200$.

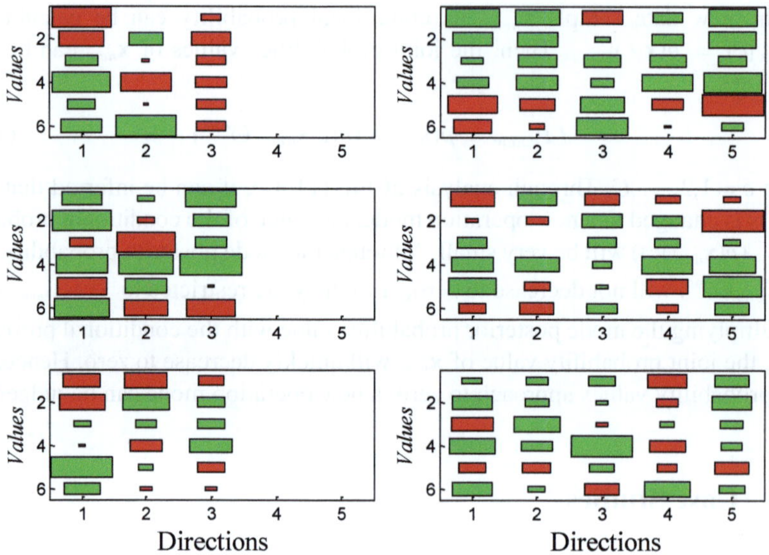

Fig. 11.1 Hinton diagrams of loading matrices of both mixture PPCA and mixture BPCA, *left* mixture BPCA; *right* mixture PPCA

First, the two mixture models are constructed with full possible latent variables direction. Denote the loading matrices of mixture PPCA as O_1, O_2, and O_3, the Hinton diagrams of the loading matrices of both models are given in Fig. 11.1. In this figure, the red rectangle represents a positive value of the element in the corresponding matrix and the green one represents a negative value. Seen from the left three subfigures, it is very clear that last two latent variable directions have been switched off by the mixture BPCA model. In contrast, the importance of the latent variable direction in mixture PPCA loading matrices cannot be easily differentiated, since the values of their elements are comparative. Therefore, through the Bayesian regularization method, the number of the effective latent variables can be determined automatically. Also, the importance of each effective latent variable direction can be determined by the value of the hyperparamter α. However, due to the inherent defect of the PPCA method, the number of effective latent variables should be determined by other methods, such as cross-validation.

To test the monitoring performance of the MBPCA method, the two generated fault cases are used. The results of the first fault by both mixture PPCA and mixture BPCA methods are shown in Fig. 11.2, in which the control limits of both monitoring statistics have been selected as 99 %. It can be seen that this fault can hardly be detected by the mixture PPCA model, since both monitoring statistics are under their control limits. In contrast, the fault can be successfully detected by the mixture BPCA model after it has been introduced into the process. Similarly, the monitoring performance of the second fault can also be improved by the MBPCA method, which is given in Fig. 11.3. Although the *SPE* statistic of both methods cannot detect

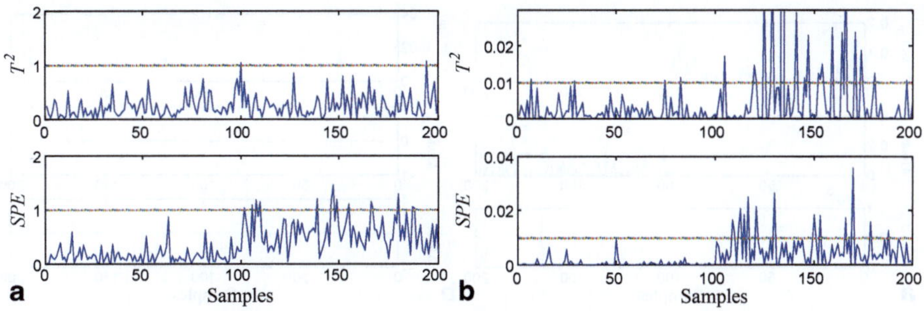

Fig. 11.2 Monitoring results of fault 1. **a** Mixture PPCA. **b** Mixture BPCA

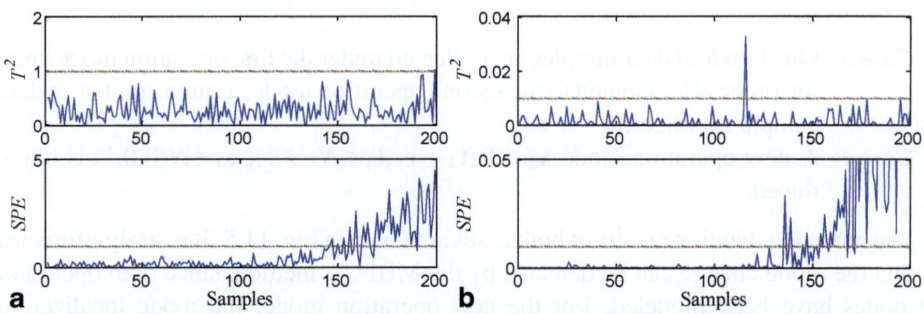

Fig. 11.3 Monitoring results of fault 2. **a** Mixture PPCA. **b** Mixture BPCA

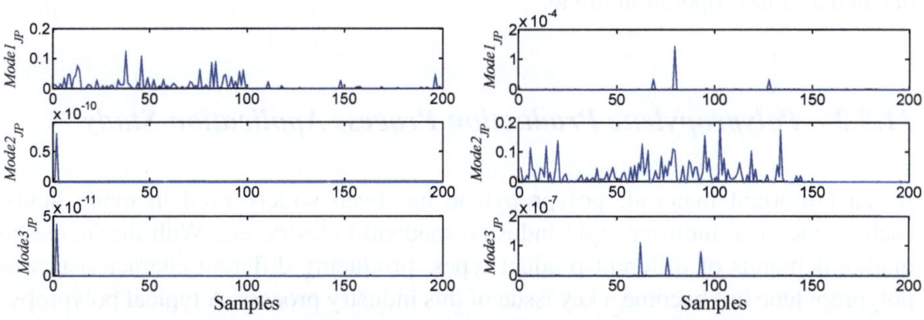

Fig. 11.4 Mode localization results. **a** The first fault. **b** The second fault

this fault, the T^2 statistic of mixture BPCA can detect the fault earlier than that of mixture BPCA. To examine the mode information of these two fault cases, their mode localization results are provided in Fig. 11.4. Judging from Fig. 11.4a, we can infer that it first ran under mode 1, and then some fault was introduced. The results presented in Fig. 11.4b exhibit that the second fault case first ran under mode 2, and then the fault happened.

In order to simulate mode change and new operation mode cases, two more datasets have also been generated, which are listed as follows:

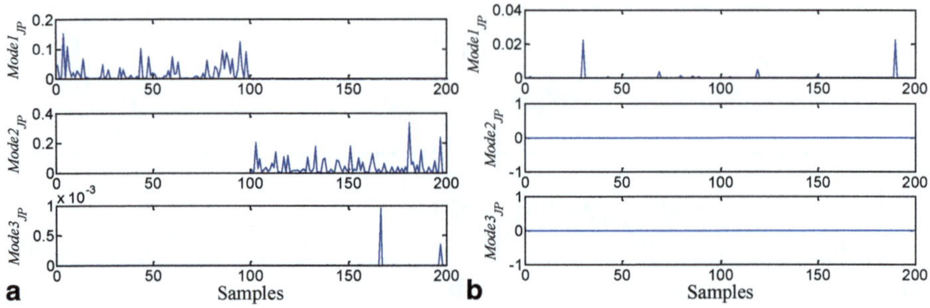

Fig. 11.5 Mode localization results of mode change and new mode cases. **a** Mode change. **b** New mode

Case 3: One hundred data samples are collected under the first operation mode, then the process is changed to the second operation mode, a same number of data sample is collected.

Case 4: A new operation mode $\mathbf{x}_4 = \mathbf{P}_4\mathbf{t}_4 + \mathbf{e}_4, \mathbf{t}_4 \in N(\text{-}3,1)$, $\mathbf{e}_4 \in N(0,0.1)$ is introduced.

The mode localization results of both cases are given in Fig. 11.5. It is straightforward that the mode change can be detected by the MBPCA method, since both operation modes have been modeled. For the new operation mode, the mode localization method can also give correct result, which can be found in Fig. 11.5b. Due to the small probability values of all the three operation modes, the new case should be deemed as a new operation mode.

11.5.2 Polypropylene Production Process Application Study

As an important material, polypropylene has been widely used in many fields, such as chemical industry, light industry, medical industry, etc. With the increased market demands of different product types, producing different characteristics of polypropylene has become a key issue of this industry process. A typical polypropylene production device always contains a catalytic body system, which comprises of $TiCl_4$, triethylaluminum (TEAL), and diphenyldimethoxysilane (DONOR). To produce different brands of productions, three reactors are connected in series. The flowchart of this process is given in Fig. 11.6. As seen in the flowchart, this process consists of four major units: the catalytic body system, and three reactors. To record the characteristic of this process, over 40 variables are measured online. However, in the present study, we have selected 14 important variables for process monitoring purpose, which are highly correlated with the final product quality. These 14 monitored variables are listed in Table 11.1.

To develop the monitoring model, three different types of industrial process datasets have been collected, each of which contains 200 data samples. Figure 11.7

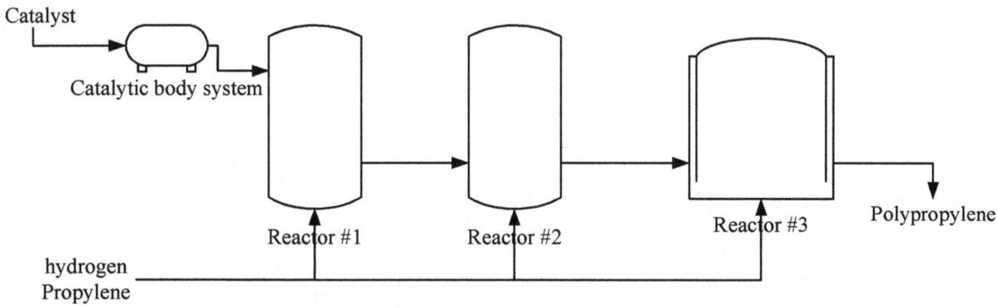

Fig. 11.6 Flowchart of the polypropylene production process

Table 11.1 Monitoring variables in polypropylene production process

No.	Measured variables	No.	Measured variables
1	Hydrogen concentration of the first reactor	8	Propylene feed of the first reactor
2	Hydrogen concentration of the second reactor	9	Propylene feed of the second reactor
3	Density of the first reactor	10	Power for the first reactor
4	Density of the second reactor	11	Power for the second reactor
5	TEAL flow	12	Lever of the second reactor
6	DONOR flow	13	Temperature of the first reactor
7	Atmer-163 flow	14	Temperature of the second reactor

exhibits the data characteristic of these three datasets, through which we can easily find that these three datasets belong to three different operation modes. To evaluate the monitoring performance of the MBPCA method, two additional fault datasets have also been collected, both of which consist of 200 data samples. These two fault cases are described as follows:

Case 1: The process initially operated under the first operation mode, 100 data samples have been collected. Then, a step change of the TEAL flow has been caused, another 100 data samples are recorded.

Case 2: The process initially operated under the third operation mode, 100 data samples have been collected. Then, a step change of the TEAL flow has been caused, another 100 data samples are recorded.

After the process datasets have been screened and preprocessed, we are ready to construct the monitoring models. Running the modeling development process 50 times, the automatically selected dimensionalities of three different subspaces are plotted in Fig. 11.8. The mean values of these three selected dimensionalities are 5.88, 6.91, and 5.94. Thus, the three dimensionalities can be determined as 6, 7, and 6 for monitoring model construction. Similarly, the Hinton diagrams of both loading matrices of the mixture PPCA and mixture BPCA models are given in Figs. 11.9 and 11.10, respectively. As expected, the dimensionality of the latent variable space can be easily determined by the mixture BPCA model, which can be clearly obtained from Fig. 11.9. However, again, the latent variable dimensionality cannot be automatically determined by the mixture PPCA model. Depending on the results

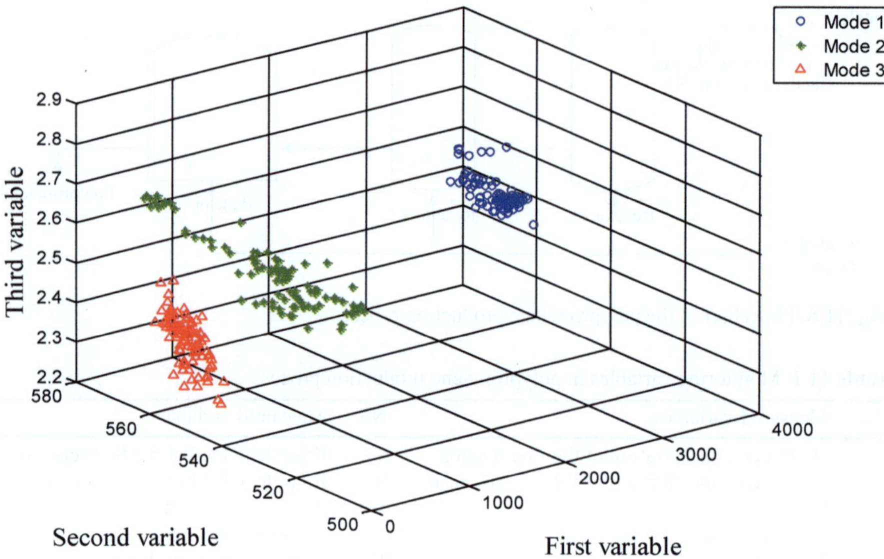

Fig. 11.7 Three-dimensional data characteristic of three different datasets in polypropylene production process

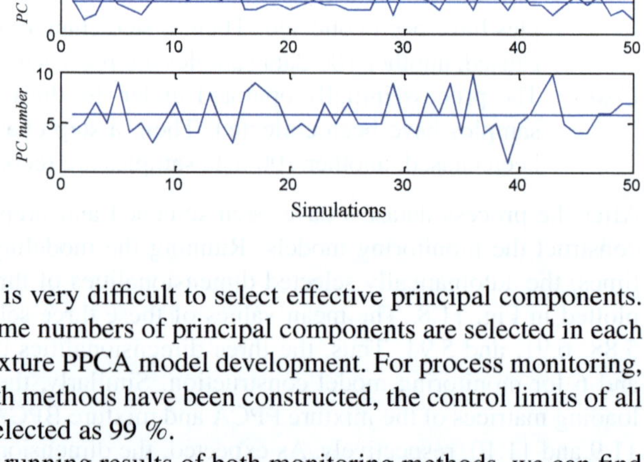

Fig. 11.8 Automatically determined number of effective latent variables in each subspace

presented in Fig. 11.10, it is very difficult to select effective principal components. However, to be fair, the same numbers of principal components are selected in each submodel space for the mixture PPCA model development. For process monitoring, T^2 and *SPE* statistic of both methods have been constructed, the control limits of all monitoring statistics are selected as 99 %.

Depending on multiple running results of both monitoring methods, we can find that the monitoring performance of both fault cases has been improved by the mixture BPCA-based method. One realization of the first fault case obtained by both methods is given in Fig. 11.11, through which one can see that both monitoring statistics of

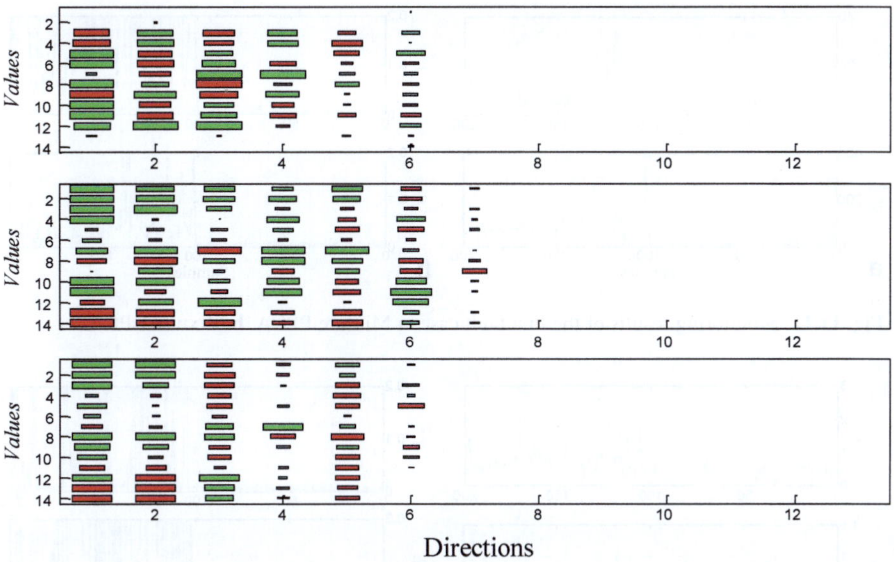

Fig. 11.9 Hinton diagrams of the loading matrix of mixture BPCA

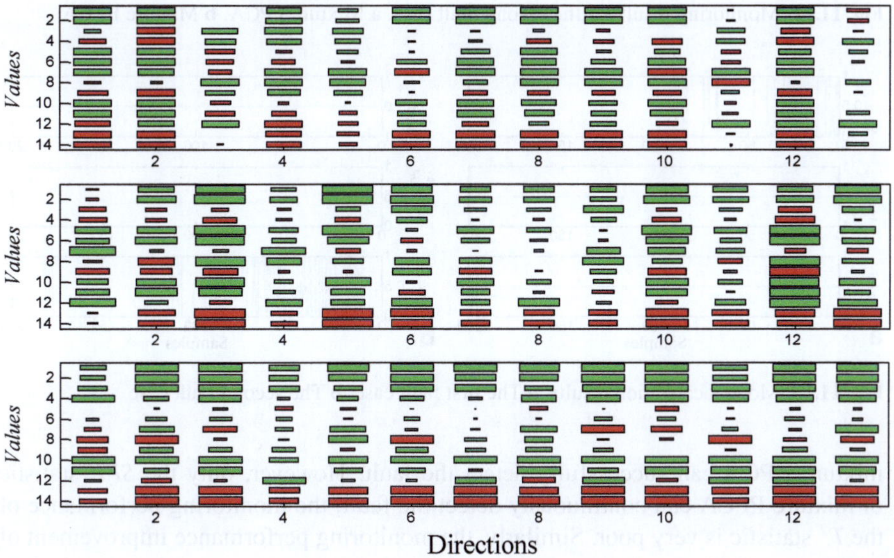

Fig. 11.10 Hinton diagrams of the loading matrix of mixture PPCA

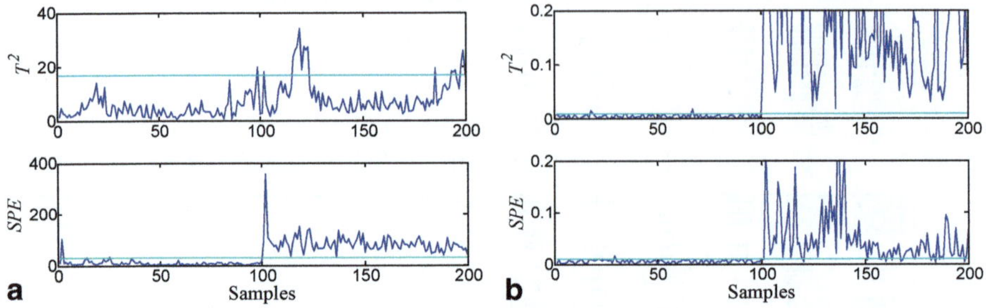

Fig. 11.11 Monitoring results of the first fault case. **a** Mixture PPCA. **b** Mixture BPCA

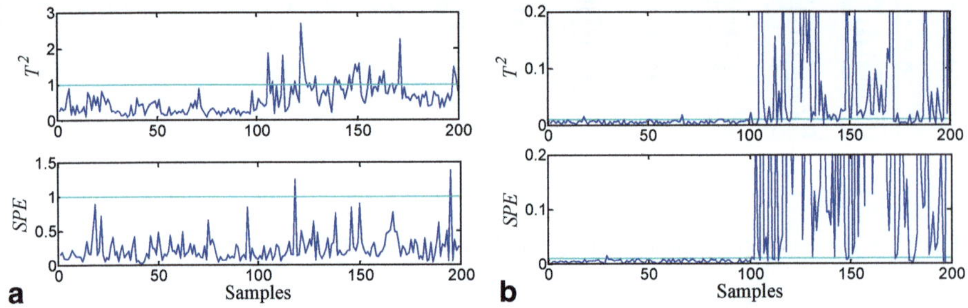

Fig. 11.12 Monitoring results of the second fault case. **a** Mixture PPCA. **b** Mixture BPCA

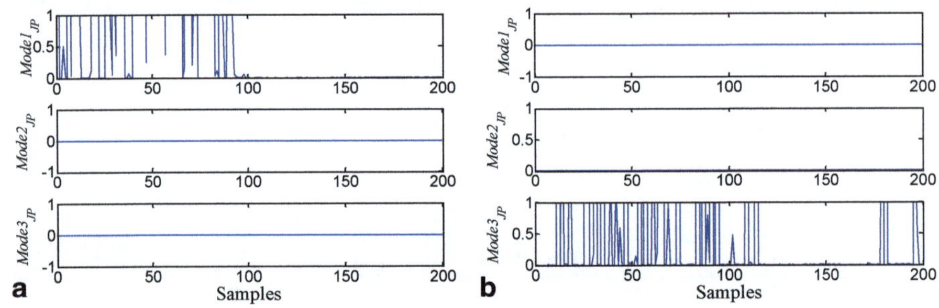

Fig. 11.13 Mode localization results. **a** The first fault case. **b** The second fault case

mixture BPCA can successfully detect the fault. However, only the *SPE* statistic of mixture PPCA can continuously detect the fault, the monitoring performance of the T^2 statistic is very poor. Similarly, the monitoring performance improvement of the mixture BPCA method case can also be obtained for the second fault, which is clearly presented in Fig. 11.12. Compared to mixture PPCA, which can hardly detect this fault, both monitoring statistics of mixture BPCA can detect it immediately after the fault happened. The mode localization results of both fault cases are shown in Fig. 11.13, depending on which one can easily conclude that these two fault cases happened under the first and the third operation modes, respectively.

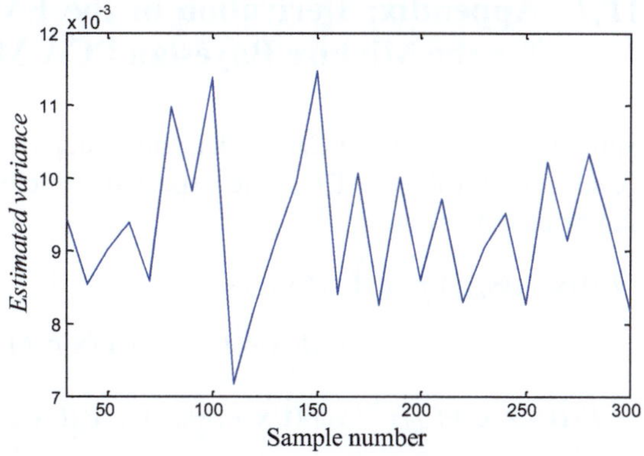

Fig. 11.14 Estimated noise variance under different numbers of data samples

In addition, the noise level of this process is examined. A total of 300 data samples of each operation mode have been collected for model development and noise variance estimation. To test the efficiency of the mixture BPCA method under limited data sample cases, we intend to change the number of modeling data samples from 30 to 300. Therefore, a total of 28 noise estimation results have been generated through the mixture BPCA modeling. These estimated results of the noise variance are shown together in Fig. 11.14, through which we can find that the modeling efficiency of the mixture BPCA method is rarely degraded when the number of data samples is changed. Therefore, when the number of process data samples is limited, which could possibly happen in multimode processes, the mixture BPCA method can still provide efficient modeling and monitoring results.

11.6 Conclusions

In the present chapter, a Bayesian regularization of the traditional PPCA method is introduced, followed by the development of the mixture Bayesian regularization method for multimode process monitoring. Compared to the traditional PPCA method, the main advantage of the Bayesian regularization method is that the effective dimensionality of the latent variable in each local space can be determined automatically. Also, it can easily determine the importance of each retained principal component, which is controlled by a hyperparameter. In addition, a mode localization approach can be easily developed in this framework, which is based on the joint probability analysis method. However, it is worth noticing that this method cannot determine if the process is operated under a new operation mode or some other changes have happened. Without appropriate process knowledge or expert experience, one can hardly obtain the exact condition of the process. How to incorporate some necessary process information into the developed monitoring scheme deserves further investigations in future researches.

11.7 Appendix: Derivation of the EM Algorithm for the Mixture Bayesian PCA Method

Similar to the single BPCA model, the E-step of the mixture BPCA model can be developed as follows. Thus, the expected sufficient statistics of the latent variable can be calculates as

$$\begin{aligned} E(\hat{\mathbf{t}}|\mathbf{x}, c, \Theta_{old}) &= E(\hat{\mathbf{t}}|c, \Theta_{old}) + \\ &\quad \text{cov}\,(\hat{\mathbf{t}}, \mathbf{x}|c, \Theta_{old})[\,\text{var}\,(\mathbf{x}|c, \Theta_{old})]^{-1}[\mathbf{x} - E(\mathbf{x}|c, \Theta_{old})] \end{aligned} \quad (11.49)$$

$$\begin{aligned} E(\hat{\mathbf{t}}\hat{\mathbf{t}}^T|\mathbf{x}, c, \Theta_{old}) &= \text{var}\,(\hat{\mathbf{t}}|\mathbf{x}, c, \Theta_{old}) + E(\hat{\mathbf{t}}|\mathbf{x}, c, \Theta_{old})E^T(\hat{\mathbf{t}}|\mathbf{x}, c, \Theta_{old}) \\ \text{var}\,(\hat{\mathbf{t}}|\mathbf{x}, c, \Theta_{old}) &= \text{var}\,(\hat{\mathbf{t}}|c, \Theta_{old}) - \\ &\quad \text{cov}\,(\hat{\mathbf{t}}, \mathbf{x}|c, \Theta_{old})[\,\text{var}\,(\mathbf{x}|c, \Theta_{old})]^{-1}\,\text{cov}\,(\mathbf{x}, \hat{\mathbf{t}}|c, \Theta_{old}) \end{aligned} \quad (11.50)$$

where $c = 1, 2, \ldots, C$, $E(\hat{\mathbf{t}}|c, \Theta_{old}) = 0$, $\text{var}\,(\mathbf{x}|c, \Theta_{old}) = \mathbf{P}_c\mathbf{P}_c^T + \beta_c^{-1}\mathbf{I}$, $E(\mathbf{x}|c, \Theta_{old}) = 0$, $\text{var}\,(\hat{\mathbf{t}}|c, \Theta_{old}) = \mathbf{I}$, and

$$\text{cov}\,(\hat{\mathbf{t}}, \mathbf{x}|c, \Theta_{old}) = E(\hat{\mathbf{t}}\mathbf{x}^T|c, \Theta_{old}) - E(\hat{\mathbf{t}}|c, \Theta_{old})E(\mathbf{x}^T|c, \Theta_{old}) = \mathbf{P}_c^T \quad (11.51)$$

$$\text{cov}\,(\mathbf{x}, \hat{\mathbf{t}}|c, \Theta_{old}) = E(\mathbf{x}\hat{\mathbf{t}}^T|c, \Theta_{old}) - E(\mathbf{x}|c, \Theta_{old})E(\hat{\mathbf{t}}^T|c, \Theta_{old}) = \mathbf{P}_c \quad (11.52)$$

Then, Eqs. (49) and (50) come to

$$E(\hat{\mathbf{t}}|\mathbf{x}, c, \Theta_{old}) = \mathbf{P}_c^T(\mathbf{P}_c\mathbf{P}_c^T + \beta_c^{-1}\mathbf{I})^{-1}\mathbf{x} = (\mathbf{P}_c^T\mathbf{P}_c + \beta_c^{-1}\mathbf{I})^{-1}\mathbf{P}_c^T\mathbf{x} \quad (11.53)$$

$$\begin{aligned} E(\hat{\mathbf{t}}\hat{\mathbf{t}}^T|\mathbf{x}, c, \Theta_{old}) &= \mathbf{I} - \mathbf{P}_c^T(\mathbf{P}_c\mathbf{P}_c^T + \beta_c^{-1}\mathbf{I})^{-1}\mathbf{P}_c + E(\hat{\mathbf{t}}|\mathbf{x}, c, \Theta_{old})E^T(\hat{\mathbf{t}}|\mathbf{x}, c, \Theta_{old}) \\ &= \beta_c^{-1}(\mathbf{P}_c^T\mathbf{P}_c + \beta_c^{-1}\mathbf{I})^{-1} + E(\hat{\mathbf{t}}|\mathbf{x}, c, \Theta_{old})E^T(\hat{\mathbf{t}}|\mathbf{x}, c, \Theta_{old}) \end{aligned} \quad (11.54)$$

Different from the EM algorithm for the single BPCA model, the proportional value of each mode $p(c|\Theta_{new})$ should also be updated through the M-step in the mixture BPCA model structure, which are calculated as follows

$$\begin{aligned} p(c|\Theta_{new}) &= \frac{1}{n}\sum_{i=1}^{n} p(c|\mathbf{x}_i, \Theta_{old}) \\ p(c|\mathbf{x}, \Theta_{old}) &= \frac{p(\mathbf{x}|c, \Theta_{old})p(c|\Theta_{old})}{p(\mathbf{x}|\Theta_{old})} \end{aligned} \quad (11.55)$$

The updated value $\mathbf{P}_c^{new}$, β_c^{new} can be determined by maximizing the log posterior probability function. Setting the derivative of $\ln p\,(\mathbf{P}|\mathbf{X})$ with respect to $\mathbf{P}_c$ and β_c as zero

$$\frac{\partial[\ln p(\mathbf{P}|\mathbf{X})]}{\partial \mathbf{P}_c} = 0 \quad (11.56)$$

$$\frac{\partial[\ln p(\mathbf{P}|\mathbf{X})]}{\partial \beta_c} = 0 \tag{11.57}$$

Then, we can easily obtain

$$\mathbf{P}_c^{new} = \left(\sum_{i=1}^{n} p(c|\mathbf{x}_i, \Theta_{old})\mathbf{x}_i E(\hat{\mathbf{t}}_i^T|\mathbf{x}_i, c, \Theta_{old})\right) \times \left(\sum_{i=1}^{n} p(c|\mathbf{x}_i, \Theta_{old})E(\hat{\mathbf{t}}_i\hat{\mathbf{t}}_i^T|\mathbf{x}_i, c, \Theta_{old}) + p(c|\Theta_{new})\beta_c^{-1}\mathbf{A}_c\right)^{-1}$$

$$\beta_c^{new} = \frac{nm}{\sum_{i=1}^{n} p(c|\mathbf{x}_i, \Theta_{old})\{\mathbf{x}_i^T\mathbf{x}_i - 2E^T(\hat{\mathbf{t}}_i|\mathbf{x}_i)\mathbf{P}_c^{newT}\mathbf{x}_i + Tr[E(\hat{\mathbf{t}}_i\hat{\mathbf{t}}_i^T|\mathbf{x}_i)\mathbf{P}_c^{newT}\mathbf{P}_c^{new}]\}} \tag{11.58}$$

Chapter 12
Plant-Wide Process Monitoring: Multiblock Method

12.1 Introduction

As a data-driven process monitoring methodology, multivariate statistical process controls (MSPCs), such as principal component analysis (PCA) and partial least squares, have been widely researched and applied to chemical plants. However, most of the modern chemical processes are always very complex, and their process variables always come from many different processing units. These processes are known as plant-wide processes. Process monitoring and diagnosis of modern plant-wide processes are complicated and the results obtained from traditional MSPC methods are always difficult to interpret. In the past years, hierarchical and multi-block approaches were developed, which divide the total variable block into several meaningful sub-blocks (MacGregor et al. 1994; Westerhuis et al. 1998; Qin and Li 2001; Smilde et al. 2003; Choi and Lee 2005; Cherry and Qin 2006). However, most of multiblock methods did not consider the cross-information between divided sub-blocks. Therefore, it is possible that a deviation of the cross-information will not be observed within any sub-blocks.

After the fault has been detected, the fault identification step should be carried out. Conventional fault identification methods include contribution plot-based methods (Westerhuis et al. 2000), fault subspace-based methods (Dunia and Qin 1998a) and so on. However, the contribution plot-based method can only narrow down the root cause of the detected fault. To determine the root cause of the fault more accurately, the reconstruction-based method can be employed (Dunia and Qin 1998b; Wang et al. 2002; Lieftuche et al. 2006). While the reconstruction methods were developed in principal component subspace (PCS) and residual subspace (RS; Dunia and Qin 1998a; Wang et al. 2002), this method was later developed in the independent component subspace (ICS; Ge and Song 2009). By employing this method, the fault subspace should be preliminarily defined. However, while it is easy to derive the fault subspace for sensor faults, the derivation of the fault subspace is not straightforward for process faults without any knowledge. To address this problem, a subspace extraction method was proposed to extract fault directions from historical fault data (Yue and Qin 2001).

Z. Ge, Z. Song, *Multivariate Statistical Process Control,* Advances in Industrial Control,
DOI 10.1007/978-1-4471-4513-4_12,

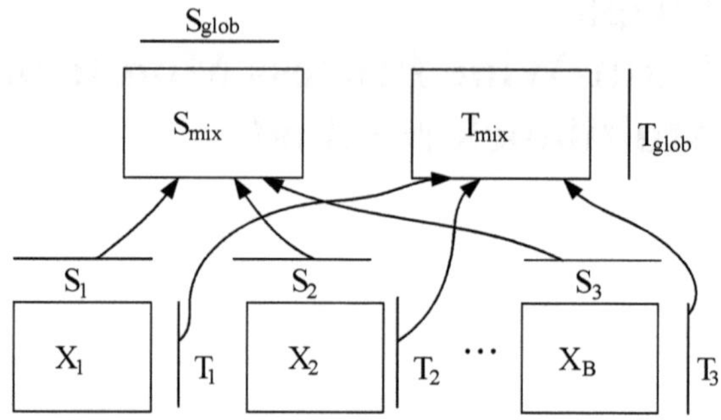

Fig. 12.1 Two-level MBICA-PCA model strategy

In this chapter, a plant-wide process monitoring strategy is introduced, which is based on two-level multiblock method and the independent component analysis and PCA (ICA-PCA) information extraction strategy. The total process variables are firstly divided into several sub-blocks, and sub-models are built correspondingly. Then, a global ICA-PCA model is built on the high level for global process monitoring. The cross-information between sub-blocks can be extracted by the high level model, and also it can enhance further dimensionality reduction for process variables. Once a fault is detected, contribution plots are first used to calculate the responsibility of each sub-block. Then, the reconstruction-based method is carried out in the responsible sub-block for advanced fault identification.

12.2 Multiblock ICA-PCA (MBICA-PCA)

Assume the data matrix $\mathbf{X} \in \mathbf{R}^{n \times m}$, where n is the number of observations and m is number of variables. According to the plant-wide process, the number of variables m is always very large. Although conventional MSPC methods show good efficiency for multicorrelated dataset, they are limited in fault diagnosis and identification. Results obtained from these conventional MSPC methods are often difficult to interpret. In order to improve fault detection and identification for large complex processes, a two-level multiblock method has been proposed (Ge and Song 2009). First of all, process variables are divided into several sub-blocks based on prior knowledge

$$\mathbf{X} = [\mathbf{X}_1 \quad \mathbf{X}_2 \quad \cdots \quad \mathbf{X}_B], \tag{12.1}$$

where B is the number of sub-blocks, each sub-block $\mathbf{X}_b \in \mathbf{R}^{n \times m_b} (b = 1, 2, \cdots, B)$ has m_b variables. The main idea of the two-level multiblock ICA-PCA method is described as follows. Firstly, a sub-model is built for each of the sub-block. Then, a global model is built on a high level, in which non-Gaussian and Gaussian information are modelled separately. The two-level multiblock ICA-PCA method is illustrated in Fig. 12.1.

In the conventional multiblock PCA algorithm (Qin and Liet al. 2001), each variable in the data block $\mathbf{X}_b$ is often scaled to have zero mean and variance of

$1/m_b$ to make each block contribute about the same variance to the super scores. For simplicity, we assume the dataset $\mathbf{X}$ has already been scaled. Therefore, B sub-models are first built for corresponding sub-blocks $\mathbf{X}_b$ ($b = 1,2, \ldots, B$).

$$\mathbf{X}_b^T = \mathbf{A}_b\mathbf{S}_b + \mathbf{E}_b, \tag{12.2}$$

$$\mathbf{E}_b = \mathbf{T}_b\mathbf{P}_b^T + \mathbf{F}_b. \tag{12.3}$$

It is important to note that some sub-blocks only contain the Gaussian information, which can be judged by negentropy (Ge and Song 2007). For these sub-blocks, PCA is enough for modelling the correlation within these sub-blocks. After all of the B sub-models have been built, the extracted independent components (ICs) and principal components (PCs) in each sub-block are arranged as follows

$$\mathbf{S}_{\text{mix}} = [\mathbf{S}_1\ \mathbf{S}_2\ \cdots\ \mathbf{S}_B]^T, \tag{12.4}$$

$$\mathbf{T}_{\text{mix}} = [\mathbf{T}_1\ \mathbf{T}_2\ \cdots\ \mathbf{T}_B]. \tag{12.5}$$

Although the correlations within each sub-block are well extracted, the correlations between sub-blocks are not well extracted. Therefore, a global model could be built on a high level to extract the cross-information, and also to model non-Gaussian and Gaussian information separately. According to the defined data matrix $\mathbf{S}_{\text{mix}}$ and $\mathbf{T}_{\text{mix}}$, the global ICA-PCA model is built as follows

$$\mathbf{S}_{\text{mix}} = \mathbf{A}_{\text{glob}}\mathbf{S}_{\text{glob}} + \mathbf{E}_{\text{glob}}, \tag{12.6}$$

$$\mathbf{T}_{\text{mix}} = \mathbf{T}_{\text{glob}}\mathbf{P}_{\text{glob}}^T + \mathbf{F}_{\text{glob}}. \tag{12.7}$$

Note that block dividing plays an important role in the multiblock method. Therefore, it is important to make intelligent selections of variable blocks based on process knowledge such that the most important correlations are extracted within divided sub-blocks. When the built model is used for process monitoring and fault diagnosis, it is much easier to isolate and identify the detected fault. Furthermore, with the development of the global model, cross-information between sub-blocks can be efficiently extracted. The variable dimension may also be reduced at the second time. Hence, more redundant information can be removed from the monitoring information, which will also enhance the performance of process monitoring.

12.3 Process Monitoring Based on Two-Level MBICA-PCA

In this section, the MBICA-PCA method is demonstrated, including fault detection and identification. Assume that we have already built all the models described in the previous section, including a global model and B sub-models. To monitor the Gaussian and non-Gaussian information separately, three statistics can be established for each ICA-PCA model. If the statistic confidence limits are exceeded, some disturbance or fault may happen in the process. Then the contribution plot of each statistic can be used to identify the responsible sub-block for the abnormal event. After that, advanced fault identification can be carried out in the responsible sub-block to find the root cause of the detected fault.

12.3.1 Fault Detection

Given the sub-block data matrix, $\mathbf{X}_b$. Assume r_b ICs are extracted, $\mathbf{S}_b = [\mathbf{s}_{b1}, \mathbf{s}_{b2}, \cdots, \mathbf{s}_{bn}] \in \mathbf{R}^{r_b \times n}$. In order to monitor the non-Gaussian part of the process, the I^2 statistic variable is defined (Lee et al. 2004)

$$I_b^2 = \mathbf{s}_b^T \mathbf{s}_b. \tag{12.8}$$

After the non-Gaussian information has been extracted, the residual matrix $\mathbf{E}_b$ is obtained. Then, we use PCA to analyze it, expanding $\mathbf{E}_b$ as below

$$\mathbf{E}_b = \sum_{i=1}^{k_b} \mathbf{t}_i \mathbf{p}_i^T + \mathbf{F}_b, \tag{12.9}$$

where k_b is the number of PCs and $\mathbf{F}_b$ is the residual resulting from the PCA model. Here, we define the limits of T_b^2 and SPE_b statistics as follows

$$T_b^2 = \sum_{i=1}^{k_b} \frac{\mathbf{t}_i \mathbf{t}_i^{\mathrm{T}}}{\lambda_i} \leq \frac{k_b(n-1)}{n-k_b} F_{k,(n-k_b),\alpha}, \tag{12.10}$$

$$\mathrm{SPE}_b = \mathbf{f}\mathbf{f}^T = \mathbf{e}(\mathbf{I} - \mathbf{P}_b \mathbf{P}_b^T)\mathbf{e}^T \leq \mathrm{SPE}_{b,\mathrm{lim}}, \tag{12.11}$$

$$\mathrm{SPE}_{b,\mathrm{lim}} = g_b \chi^2_{h_b,\alpha}, \tag{12.12}$$

where $g_b = v_b/(2m_b)$, $h_b = 2m_b^2/v_b$, m_b and v_b are the mean and variance values of squared prediction error (SPE) in each of the sub-blocks, k_b is the number of PCs, α is the significance level and c_α is the normal deviate corresponding to the upper $1-\alpha$ percentile.

On the high level of the model, the monitoring data matrices become the new arranged data matrices $\mathbf{S}_{\mathrm{mix}}$ and $\mathbf{T}_{\mathrm{mix}}$, similar statistics can be established. Assume $\mathbf{S}_{\mathrm{glob}}$ is the IC matrix calculated by the global ICA model and $\mathbf{T}_{\mathrm{glob}}$ is the score matrix of $\mathbf{T}_{\mathrm{mix}}$ calculated by the global PCA model. The three statistics are established as follows

$$I_{\mathrm{glob}}^2 = \mathbf{S}_{\mathrm{glob}}^T \mathbf{S}_{\mathrm{glob}}, \tag{12.13}$$

$$T_{\mathrm{glob}}^2 = \mathbf{t}_{\mathrm{glob}} \Lambda_{\mathrm{glob}} \mathbf{t}_{\mathrm{glob}}^T, \tag{12.14}$$

$$\mathrm{SPE}_{\mathrm{glob}} = \mathbf{f}_{\mathrm{glob}}\mathbf{f}_{\mathrm{glob}}^T. \tag{12.15}$$

In PCA monitoring, the confidence limits are based on a specified distribution shown in Eqs. 12.10–12.12 based upon the assumption that the latent variables follow a Gaussian distribution. However, in ICA monitoring, the IC does not conform to a specific distribution. Hence, the confidence limit of the I^2 statistic cannot be determined directly from a particular approximate distribution. An alternative approach to define the nominal operating region of the I^2 statistic is to use either the kernel density estimation or the support vector description method (Chen et al. 2000; Ge and Song 2007; Ge et al. 2009). For simplicity, the confidence limit of I^2 can also be determined by try and error.

12.3.2 Fault Identification

After a fault has been detected, the contribution plot method is firstly selected to identify the responsible sub-block. For advanced fault identification, the reconstruction-based method is carried out in each of the three subspaces: ICS, PCS and RS. Therefore, the root cause of the detected fault can be further identified, hence improving the fault interpretation. In order to determine which block takes the responsibility for the fault, block contributions are defined as follows, which are average values of the variable contributions in corresponding blocks (Ge and Song 2009).

$$I^2\mathrm{Cont}_{\mathrm{block},b} = \sum_{i \in VI(b)} I^2\mathrm{Cont}_{\mathrm{glob},i}/r_b (b = 1, 2, \cdots, B), \tag{12.16}$$

$$T^2\mathrm{Cont}_{\mathrm{block},b} = \sum_{i \in VP(b)} T^2\mathrm{Cont}_{\mathrm{glob},i}/k_b (b = 1, 2, \cdots, B), \tag{12.17}$$

$$\mathrm{SPECont}_{\mathrm{block},b} = \sum_{i \in VP(b)} \mathrm{SPECont}_{\mathrm{glob},i}/k_b (b = 1, 2, \cdots, B), \tag{12.18}$$

where r_b is the number of IC extracted from block b, k_b is the number of PC extracted from block b, $VI(b)$ is the IC number extracted in the bth sub-block and $VP(b)$ is the PC number extracted in the bth sub-block. The contributions of the ith component in Eqs. 12.16–12.18 are defined as (Lee et al. 2004; Westerhuis et al. 2000)

$$I^2\mathrm{Cont}_{\mathrm{glob},i} = \frac{\mathbf{A}_{\mathrm{glob}}\mathbf{S}_{\mathrm{fault,glob},i}}{\left\|\mathbf{A}_{\mathrm{glob}}\mathbf{S}_{\mathrm{fault,glob},i}\right\|_2}\left\|\mathbf{S}_{\mathrm{fault,glob},i}\right\|_2, \tag{12.19}$$

$$T^2\mathrm{Cont}_{\mathrm{glob},i} = \sum_{i=1}^{k} \frac{\mathbf{t}_i}{\lambda_i}\mathbf{P}_{\mathrm{glob},i}\mathbf{e}_{\mathrm{fault,glob}}, \tag{12.20}$$

$$\mathrm{SPECont}_{\mathrm{glob},i} = \left(\mathbf{f}_{\mathrm{fault,glob},i} - \hat{\mathbf{f}}_{\mathrm{fault,glob},i}\right)^2. \tag{12.21}$$

Here, the fault subspace is supposed to be known. Otherwise, it can be extracted by the singular vector decomposition method proposed by Yue and Qin (2001). Assume the fault set includes J faults, which is described as $\{\mathbb{F}_j, j=1, 2, \ldots, J\}$. Denote $\{\Theta_j, j=1, 2, \ldots, J\}$ as the fault subspace with dimensions $\{\dim(\Theta_j) \geq 1, j = 1, 2, \ldots, J\}$ for the defined fault set. Hence, both of the unidimensional and multidimensional faults are considered.

After the fault subspace has been defined, the normal value $\mathbf{x}^*$ can be reconstructed from the corrupted value $\mathbf{x}$ in each of the IC, PC and RSs. Assume a fault $\mathbb{F}_j$ has happened, a reconstructed value x_j can be obtained as an adjustment of the corrupted value $\mathbf{x}$ moving along a given fault direction $\boldsymbol{\Theta}_j$

$$\mathbf{x}_j = \mathbf{x} - \Theta_j \mathbf{f}_j, \tag{12.22}$$

where $\mathbf{f}_j$ is the estimated fault vector with $\|\mathbf{f}\|$ as its magnitude such that $\mathbf{x}_j$ is closest to the normal region.

In the IC subspace, theoretically, the optimal reconstruction is obtained by minimizing $\left\|\mathbf{x}_j - \mathbf{x}^*\right\|$. However, this is infeasible because the normal value $\mathbf{x}^*$ is unknown. Alternatively, the reconstruction is realized by moving the corrupted value $\mathbf{x}$ along the defined fault direction $\boldsymbol{\Theta}_j$ in ICS. The reconstruction can be formulized in the following optimization problem (Ge and Song 2009)

$$\mathbf{f}_j = \arg\min_{\mathbf{f}} \left\|\mathbf{W}(\mathbf{x} - \Theta_j \mathbf{f})\right\|^2. \tag{12.23}$$

Denote $\Delta_j = \mathbf{W}\boldsymbol{\Theta}_j$, the solution of Eq. (12.23) is straightforward by least square technique, which yields as follows

$$\mathbf{f}_j = (\boldsymbol{\Delta}_j^T \boldsymbol{\Delta}_j)^{-1} \Delta_j^T \mathbf{W}\mathbf{x}, \tag{12.24}$$

where the column rank of the matrix Δ_j is assumed to be full. If the matrix Δ_j suffers a column rank deficiency, then the corresponding solution of the optimization problem in Eq. (12.23) will be

$$\mathbf{f}_j = \Delta_j^+ \mathbf{W}\mathbf{x}, \tag{12.25}$$

where the matrix Δ_j^+ is the Moore–Penrose pseudo-inverse of Δ_j. Hence, the reconstructed data vector can be represented as Eq. (12.22). The new I^2 statistic value of this reconstructed data sample can be calculated as follows

$$I_j^2 = \mathbf{S}_j^T \mathbf{S}_j, \tag{12.26}$$

where $\mathbf{S}_j = \mathbf{W}\mathbf{x}_j$, then the identification index in ICS is defined as

$$\eta_{\mathrm{ICS},j} = \frac{I_j^2}{I_{\mathrm{fault}}^2}, \tag{12.27}$$

where I_{fault}^2 is the value of statistic with faulty data. Therefore, when the fault subspace $\boldsymbol{\Theta}_j$ is matched, the value of $\eta_{\mathrm{ICS},j}$ reduces significantly.

Similarly, the detected fault can be reconstructed in both PCS and RS. Related works have been published, including Dunia and Qin (1998a), Wang et al. (2002) and Lieftuche et al. (2006). Therefore, the T^2 and SPE statistic values of the reconstructed data sample can be calculated, respectively, in PCS and RS

$$T_j^2 = \left\| \boldsymbol{\lambda}^{-1/2} \mathbf{P}^{\mathrm{T}} \mathbf{x}_j \right\|^2, \tag{12.28}$$

$$\mathrm{SPE}_j = \left\| (\mathbf{I} - \mathbf{P}\mathbf{P}^{\mathrm{T}}) \mathbf{x}_j \right\|^2, \tag{12.29}$$

where $\boldsymbol{\lambda}$ is the eigenvalue matrix and $\mathbf{I}$ is an identity matrix with appropriate dimension. Then, the identification index in PC and RSs can be defined as

$$\eta_{\mathrm{PCS},j} = \frac{T_j^2}{T_{\mathrm{fault}}^2}, \tag{12.30}$$

$$\eta_{\mathrm{RS},j} = \frac{\mathrm{SPE}_j}{\mathrm{SPE}_{\mathrm{fault}}}. \tag{12.31}$$

In summary, two steps are taken for fault identification. Firstly, the contribution plot method is used to identify the responsible sub-block for the detected fault. Although this method is hard to pinpoint the specific cause of the fault, it can narrow down the possible causes to a specific sub-block or several sub-blocks. After the responsible sub-block has been determined, fault reconstruction and identification can be carried out in any of the three subspaces (ICS, PCS and RS) as long as it can be detected by its corresponding statistic. Therefore, the advanced fault identification method determines the exact root cause of the fault, and reconstructs the faulty sample value simultaneously.

The procedures and flowchart of online fault detection and identification strategy are given as follows and in Fig. 12.2.

Step 1 Scale the new sample data;
Step 2 Calculate the global model statistic values for the new data;
Step 3 If any of the statistic is beyond its limit, a fault is detected;
Step 4 Generate the global contribution plots to identify the responsible sub-block;
Step 5 Calculate values of sub-model statistics and examine if any of the statistic limit is exceeded;
Step 6 Advanced fault identification based on the reconstruction method in ICS, PCS or RS subspace.

12.4 Case Study of TE Benchmark Process

In this section, the MBICA-PCA method is tested through the Tennessee Eastman (TE) process (Downs and Vogel 1993; Chiang et al. 2001). As a benchmark simulation, the TE process has been widely used to test the performance of various

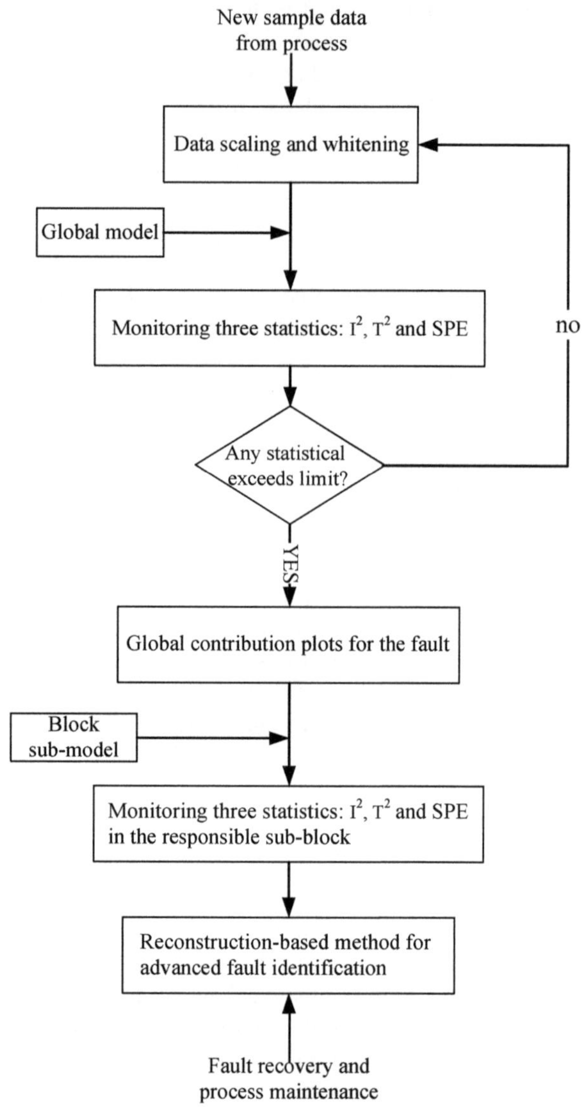

Fig. 12.2 Fault detection and identification based on MBICA-PCA

monitoring approaches. Here, it is used as an example of the plant-wide process. As we know, this process has 41 measured variables (22 continuous process measurements and 19 composition measurements) and 12 manipulated variables, and a set of 21 programmed faults are introduced to the process. The details on the process description are well explained in a book of Chiang et al. (2001) and in the previous chapters. For plant-wide process monitoring, a total of 33 variables are selected, which are also listed in Table 3.1. The simulation data which we have collected are separated into two parts: training datasets and test datasets, they both consisted of

Table 12.1 JB test results of ICs and PCs

Components	IC_{glob4}	IC_{glob5}	IC_{glob6}	IC_{glob7}	PC_{glob1}	PC_{glob2}	PC_{glob3}	PC_{glob4}
H	1	1	1	1	0	0	0	0
Components	PC_1	PC_2	PC_3	PC_4	PC_5	PC_6	PC_7	PC_8
H	1	1	1	1	1	1	1	1

960 observations for each operation mode, and their sampling interval was 3 min. All process faults are introduced in the process after sample 160.

Previously, it was logical to divide the process into sub-blocks that describe a unit or a specific physical or chemical operation. As the TE process consists of five major units, we can divide the process variables into five sub-blocks, each block corresponding to one unit. However, as the condenser unit and the compressor unit have only two variables, they are suggested to be integrated into other three sub-blocks. Therefore, these 33 monitoring variables are divided into three sub-blocks. There are 16 (1–9, 21, 23–26, 32, 33), 10 (10–14, 20, 22, 27–29) and 7 (15–19, 30, 31) variables in these three sub-blocks. To build the sub-model in the first sub-block, four ICs and six PCs are selected. The other two selections of the corresponding sub-models are three ICs, four PCs and two ICs, three PCs. On the high level modelling, seven ICs and ten PCs are selected. To examine the rationality of our choices of component numbers, Jarque-Bera (JB) test is introduced. Actually, the *jbtest* function in MATLAB toolbox is used. If the value of the *jbtest* function equals to 1, it means that the tested component is non-Gaussian. On the other hand, if the value is zero, the tested component is considered to be Gaussian. Results of JB test are tabulated in Table 12.1. For comparison, PCs of conventional PCA are also tested. Results of the last four ICs and the first four PCs of the global model are shown in the second row of Table 12.1, and results of the first eight PCs of the PCA model are given in the last row of the table. As shown in the table, the values of IC_{glob7} and PC_{glob1} are 1 and 0, it means that the non-Gaussian information is approximately extracted by seven ICs. However, the first eight PCs of PCA are all considered non-Gaussian. In fact, if 19 PCs are chosen for the PCA model, among which the first 12 PCs are considered to be non-Gaussian, only 7 PCs (13–19) are Gaussian-distributed.

To evaluate the monitoring performance of the MBICA-PCA method, a sensor fault is first simulated in the TE process. A constant value is added to variable 5, and 960 samples are collected. Global process monitoring results of this sensor fault are given in Fig. 12.3a. As indicated, all three statistic confidence limits have been exceeded, which means that a fault has been detected. To determine which sub-block takes the most responsibility of the fault, the contribution plot method is employed. It is clearly shown in Fig. 12.3b that the first sub-block should be the most responsible one for the detected fault. The results of contribution plots are straightforward and easy to be interpreted, as the faulty sensor is included in the first sub-block. Monitoring results of three sub-blocks are given in Fig. 12.4a. It can be inferred that this sensor fault has happened in the first sub-block, as most of the statistic values exceed their corresponding control limits. However, the statistic confidence limits in Fig. 12.4b, c are not violated during the operating process.

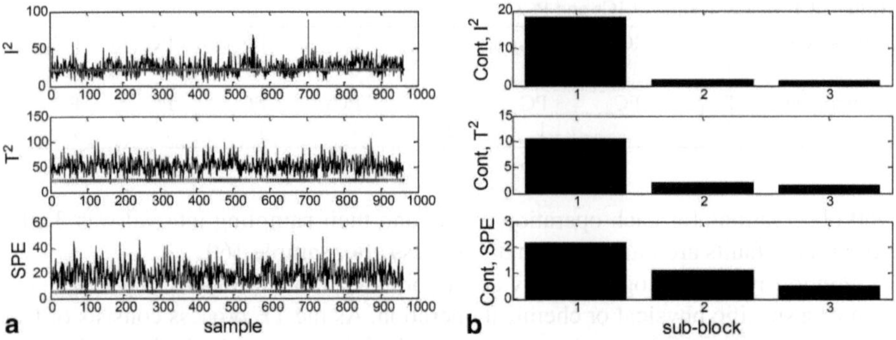

Fig. 12.3 Monitoring results of the sensor fault. **a** Fault detection, **b** Sub-block identification

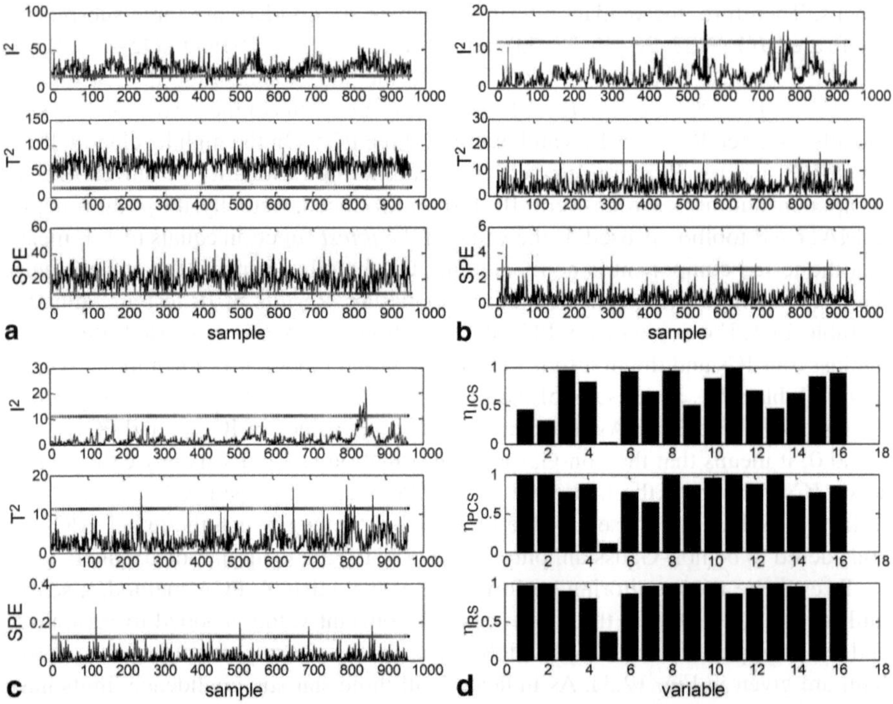

Fig. 12.4 Advanced monitoring results of the sensor fault. **a** Fault detection results of sub-block 1, **b** Fault detection results of sub-block 2, **c** Fault detection results of sub-block 3, **d** Advanced fault identification results of the sensor fault

In order to gain further interpretation and determine the exact root cause of the fault, advanced fault identification can be carried out. Thus, the reconstruction-based method is employed in the first sub-block. Final identification results are shown in Fig. 12.4d. One can find that the faulty variable (variable 5) is correctly identified in

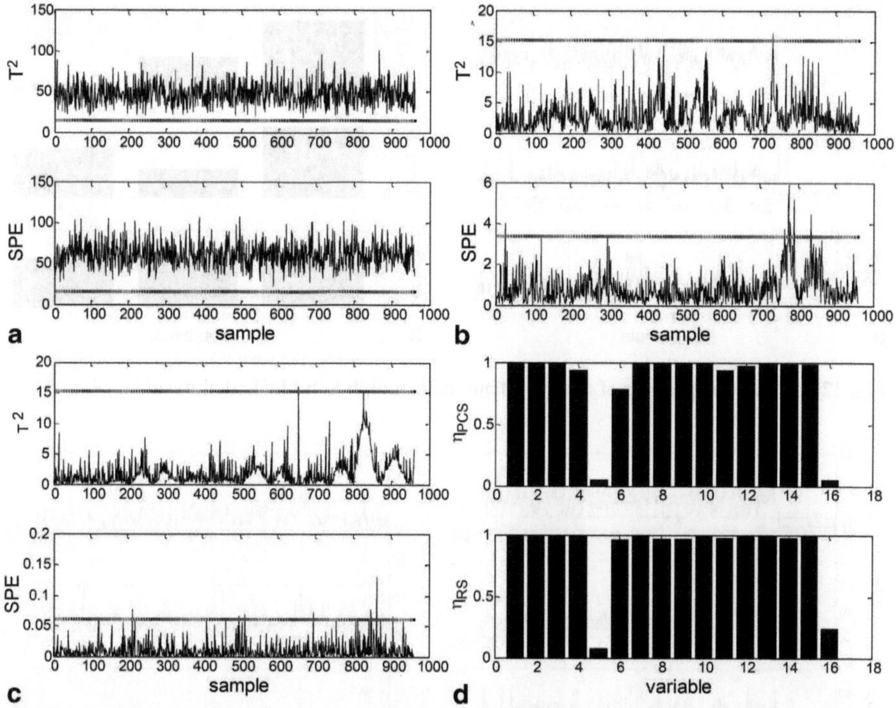

Fig. 12.5 Monitoring results of the sensor fault by multiblock PCA. **a** Fault detection results of sub-block 1, **b** Fault detection results of sub-block 2, **c** Fault detection results of sub-block 3, **d** Fault identification results of the sensor fault

any of the three subspaces (ICS, PCS and RS). For comparison, monitoring results of multiblock PCA are given in Fig. 12.5. As no super model has been built for global process monitoring, three sub-blocks should be monitored separately. If there are many sub-blocks, the monitoring tasks will become troublesome. However, if the two-level multiblock model is used, one can first use the global model to monitor the whole process, and then the advanced process monitoring should be carried out only in the responsible sub-block. Although the fault detection and identification results of the two methods seem to be identical in this case, the two-level multiblock method has more potential for monitoring plant-wide processes.

Next, fault 4 is considered. Unlike the sensor fault, whose fault subspace is easily to be derived, fault subspace derivations of process faults are not straightforward. To this end, Yue and Qin (2001) proposed a fault subspace extraction strategy for the reconstruction-based fault identification method. Before carrying out fault identification, all of the 21 fault subspaces in the TE process are supposed to be extracted and preserved. Fault detection and identification results of both methods for fault 4 are shown in Figs. 12.6, 12.7 and 12.8. As seen in Fig. 12.6, all of the three monitoring statistics indicate that a fault has happened, and the contribution plots given

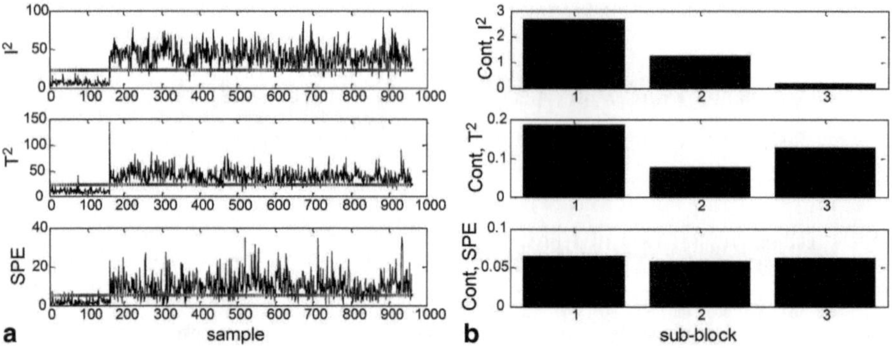

Fig. 12.6 Monitoring results of fault 4. **a** Fault detection, **b** Sub-block identification

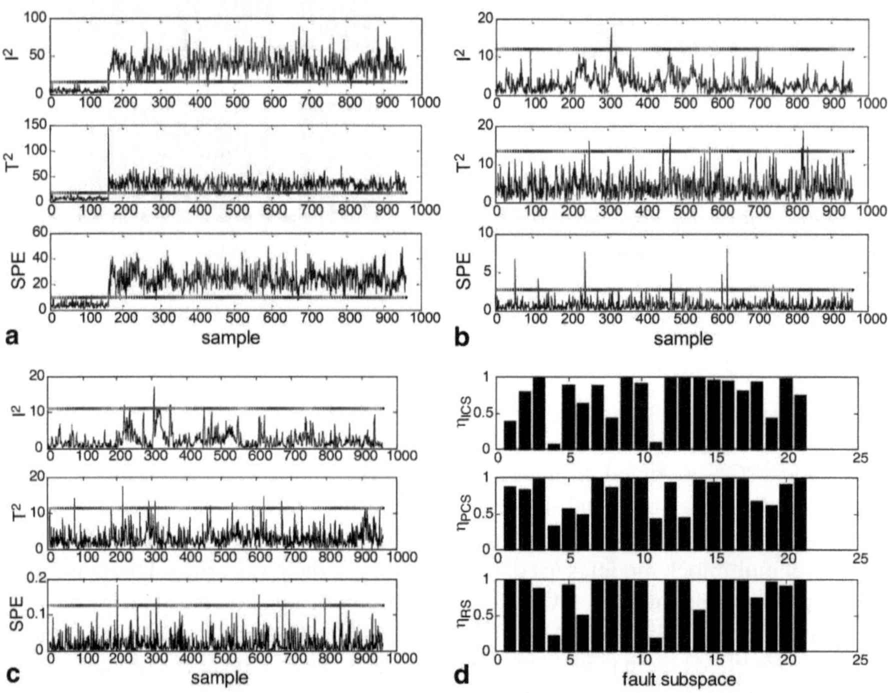

Fig. 12.7 Advanced monitoring results of fault 4. **a** Fault detection results of sub-block 1, **b** Fault detection results of sub-block 2, **c** Fault detection results of sub-block 3, **d** Advanced fault identification results of fault 4

in Fig. 12.6b indicated that the first sub-block takes the most responsibility. Therefore, the fist sub-block is further monitored and advanced fault identification is also carried out in this sub-block. To illustrate the responsibility of the first sub-block, fault detection results of the second and third sub-blocks are also given in Fig. 12.7.

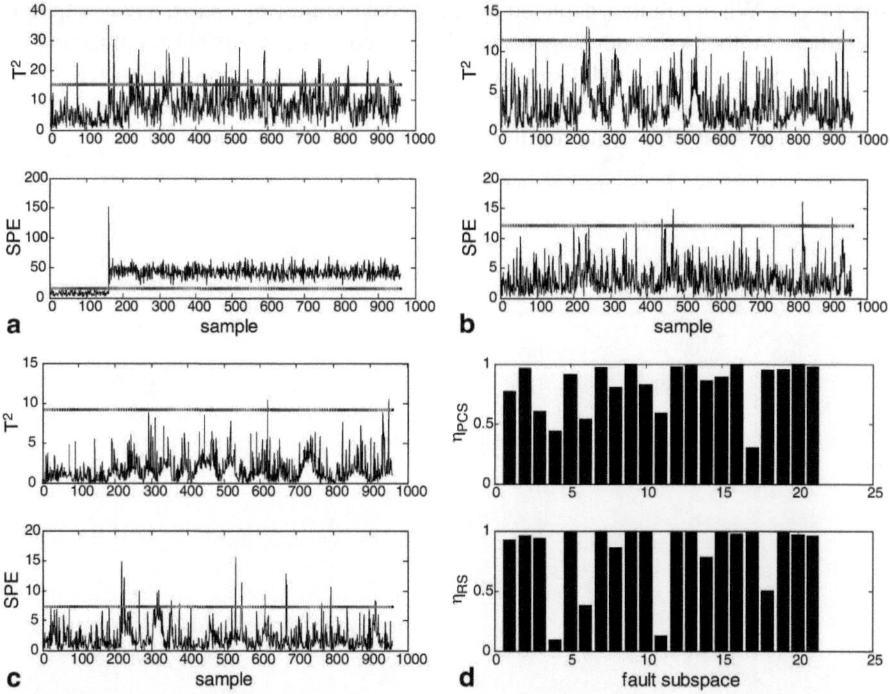

Fig. 12.8 Monitoring results of fault 4 by multiblock PCA. **a** Fault detection results of sub-block 1, **b** Fault detection results of sub-block 2, **c** Fault detection results of sub-block 3, **d** Fault identification results of fault 4

Besides, the results of advanced fault identification are also given in Fig. 12.7d, in which the fourth fault subspace is found to be the most likely one that could happen in the process. However, the 11th fault subspace seems to be similar with the identified fault subspace, which puzzles the fault identification result. In this case, the performance could be improved by incorporating some process or expert knowledge. Similar fault detection results are obtained by multiblock PCA, which are given in Fig. 12.8. However, the fault identification results seem to be worse than that in Fig. 12.6d.

12.5 Conclusions

In this chapter, a two-level MBICA-PCA method has been introduced for plant-wide process monitoring. The multiblock method first divides the process into several sub-blocks. Sub-models are built for these divided sub-blocks. Then, the extracted information from these sub-blocks is integrated on the high level. A global monitoring model is developed on this high level to extract further cross-information between the sub-blocks. When a fault is detected by the global monitoring charts, a two-step fault identification method is developed, which is based on the contribution plots and the fault reconstruction method. Compared to the conventional approach, the two-level MBICA-PCA method is more efficient for both fault detection and identification. The TE case study shows the feasibility and efficiency of this method.

References

Albazzaz H, Wang XZ (2007) Introduction of dynamics to an approach for batch process monitoring using independent component analysis. Chem Eng Commun 194:218–233

Alcala C, Qin SJ (2009) Reconstruction-based contribution for process monitoring. Automatica 45:1593–1600

AlGhazzawi A, Lennox B (2008) Monitoring a complex refining process using multivariate statistics. Control Eng Pract 16:294–307

Bakshi BR (1998) Multiscale PCA with applications to multivariate statistical process monitoring. AIChE J 40:1596–1610

Bartelmus W, Zimroz R (2009) Vibration condition monitoring of planetary gearbox under varying external load. Mech Syst Signal Process 23:246–257

Basseville M (1998) On-board component fault detection and isolation using the statistical local approach. Automatica 34:1391–1415

Basseville M, Nikiforov I (1993) Detection of abrupt changes: theory and application. Prentice Hall, New Jersey

Basseville M, Abdelghani M, Benveniste A (2000) Subspace-based fault detection algorithms for vibration monitoring. Automatica 36:101–109

Bauer D, Deistler M, Scherrer W (1999) Consistency and asymptotic normality of some subspace algorithms for systems without observed inputs. Automatica 35:1243–1254

Baydar N, Ball AD (2001) A comparative study of acoustics and vibration signals in detection of gear failures using Wigner-Ville distribution. Mech Syst Signal Process 15:1091–1107

Baydar N, Ball AD, Kruger U (1999) Detection of incipient tooth defect in helical gears using principal components. In: Proceedings of the 1st International Conference on the Integrating of Dynamics, Monitoring and Control, Manchester, pp 93–100

Baydar N, Chen Q, Ball AD, Kruger U (2001) Detection of incipient tooth defect in helical gears using multivariate statistics. Mech Syst Signal Process 15:303–321

Bhagwat A, Srinivasan R, Krishnaswamy PR (2003a) Fault detection during process transitions: a model-based approach. Chem Eng Sci 58:309–325

Bhagwat A, Srinivasan R, Krishnaswamy PR (2003b) Multi-linear model-based fault detection during process transitions. Chem Eng Sci 58:1649–1670

Bontempi G, Bersini H, Birattari M (2001) The local paradigm for modeling and control: from neuro-fuzzy to lazy learning. Fuzzy Sets Syst 121:59–72

Bicego M, Figueiredo MAT (2009) Soft clustering using weighted one-class support vector machines. Pattern Recogn 42:27–32

Bishop CM (1999) Bayesian PCA. Adv Neural Inf Process Syst 11:382–388

Bishop CM (2006) Pattern recognition and machine learning. Springer, Heidelberg

Breiman L (1996) Bagging predictors. Mach Learn 24:123–140

Chen JH, Chen HH (2006) On-line batch process monitoring using MHMT-based MPCA. Chem Eng Sci 61:3223–3239

Z. Ge, Z. Song, *Multivariate Statistical Process Control,* Advances in Industrial Control, 183
DOI 10.1007/978-1-4471-4513-4, © Springer-Verlag London 2013

Chen J, Liu J (1999) Mixture principal component analysis models for process monitoring. Ind Eng Chem Res 38:1478–1488
Chen J, Liu J (2000) Using mixture principal component analysis networks to extract fuzzy rules from data. Ind Eng Chem Res 39:2355–2367
Chen J, Liu K (2002) On-line batch process monitoring using dynamic PCA and dynamic PLS models. Chem Eng Sci 14:63–75
Chen T, Sun Y (2009) Probabilistic contribution analysis for statistical process monitoring: a missing variable approach. Control Eng Pract 17:469–477
Chen WC, Wang MS (2009) A fuzzy c-means clustering-based fragile watermarking scheme for image authentication. Expert Syst Appl 36:1300–1307
Chen T, Zhang J (2010) On-line multivariate statistical monitoring of batch processes using Gaussian mixture model. Comput Chem Eng 34:500–507
Chen Q, Wynne RJ, Goulding P, Sandoz D (2000) The application of principal component analysis and kernel density estimation to enhance process monitoring. Control Eng Pract 8:531–543
Chen Q, Kruger U, Leung ATY (2004) Regularised kernel density estimation for clustered process data. Control Eng Pract 12:267–274
Cheng C, Chiu MS (2004) A new data-based methodology for nonlinear process modeling. Chem Eng Sci 59:2801–2810
Cheng C, Chiu MS (2005) Nonlinear process monitoring using JITL-PCA. Chemometr Intell Lab Syst 76:1–13
Cherry GA, Qin SJ (2006) Multiblock principal component analysis based on a combined index for semiconductor fault detection and diagnosis. IEEE Trans Semiconductor Manuf 19:159–172
Chiang LH, Russell EL, Braatz RD (2001) Fault detection and diagnosis in industrial systems. Springer, London
Cho JH, Lee JM, Choi SW, Lee DW, Lee IB (2005) Fault identification for process monitoring using kernel principal component analysis. Chem Eng Sci 60:279–288
Choi SW, Lee IB (2005) Multiblock PLS-based localized process diagnosis. J Process Control 15:295–306
Choi SW, Park JH, Lee IB (2004) Process monitoring using a Gaussian mixture model via principal component analysis and discriminant analysis. Comput Chem Eng 28:1377–1387
Choi SW, Lee CK, Lee JM, Park JH, Lee IB (2005) Fault detection and identification of nonlinear processes based on kernel PCA. Chemometr Intell Lab Syst 75:55–67
Dayal BS, MacGregor JF (1997) Recursive exponentially weighted PLS and its applications to adaptive control and prediction. J Process Control 7:169–179
Dong D, McAvoy TJ (1996) Nonlinear principal component analysis-based on principal curves and neural networks. Comput Chem Eng 20:65–78
Downs JJ, Vogel EF (1993) A plant-wide industrial process control problem. Comput Chem Eng 17:245–255
Dunia R, Qin SJ (1998a) Subspace approach to multidimensional fault identification and reconstruction. AIChE J 44:1813–1831
Dunia R, Qin SJ (1998b) A unified geometric approach to process and sensor fault identification: the unidimensional fault case. Comput Chem Eng 22:927–943
Eftekharnejad B, Mba D (2008) Acoustic emission signals associated with damaged helical gears. Insight Non-Destr Test Cond Monit 50:450–453
Fugate ML, Sohn H, Farrar CR (2001) Vibration-based damage detection using statistical process control. Mech Syst Signal Process 15:707–721
Ge ZQ, Song ZH (2007) Process monitoring based on independent component analysis-principal component analysis (ICA-PCA) and similarity factors. Ind and Eng Chem Res 46:2054–2063
Ge ZQ, Song ZH (2008) Online monitoring of nonlinear multiple mode processes based on adaptive local model approach. Control Eng Pract 16:1427–1437
Ge ZQ, Song ZH (2009a) Two-level multiblock statistical monitoring for plant-wide processes. Korean J Chem Eng 26: 1467–1475

Ge ZQ, Song ZH (2009b) Multimode process monitoring based on Bayesian method. J Chemometr 23:636–650
Ge ZQ, Song ZH (2010a) Maximum-likelihood mixture factor analysis model and its application for process monitoring. Chemometr Intell Lab Syst 102:53–61
Ge ZQ, Song ZH (2010b) Mixture Bayesian regularization method of PPCA for multimode process monitoring. AIChE J 56:2838–2849
Ge ZQ, Yang CJ, Song ZH, Wang HQ (2008) Robust online monitoring for multimode processes based on nonlinear external analysis. Ind Eng Chem Res 47:4775–4783
Ge ZQ, Xie L, Song ZH (2009a) A novel statistical-based monitoring approach for complex multivariate processes. Ind Eng Chem Res 48:4892–4898
Ge, ZQ, Yang CJ, Song ZH (2009b) Improved kernel PCA-based monitoring approach for nonlinear processes. Chem Eng Sci 64:2245–2255
Ge ZQ, Xie L, Kruger U, Lamont L, Song ZH, Wang SQ (2009c) Sensor fault identification and isolation for multivariate non-Gaussian processes. J Process Control 19:1707–1715
Ge ZQ, Kruger U, Lamont L, Xie L, Song ZH (2010a) Fault detection in non-Gaussian vibration systems using dynamic statistical-based approaches. Mech Syst Signal Process 24:2972–2984
Ge ZQ, Zhang MG, Song ZH (2010b) Nonlinear process monitoring based on linear subspace and Bayesian inference. J Process Control 20:676–688
Ge ZQ, Gao FR, Song ZH (2011) Two-dimensional Bayesian monitoring method for nonlinear multimode processes. Chem Eng Sci 66:5173–5183
Gertler J, Li WH, Huang YB, McAvoy T (1999) Isolation enhanced principal component analysis. AIChE J 45:323–334
He QB, Kong FR, Yan RQ (2007a) Subspace-based gearbox condition monitoring by kernel principal component analysis. Mech Syst Signal Process 21:1755–1772
He QB, Feng ZH, Kong FR (2007b) Detection of signal transients using independent component analysis and its application in gearbox condition monitoring. Mech Syst Signal Process 21:2056–2071
He QB, Yan RQ, Kong FR, Du RX (2009) Machine condition monitoring using principal component representations. Mech Syst Signal Process 23:446–466
Hiden HG, Willis MJ, Tham MT, Montague GA (1999) Non-linear principal component analysis using genetic programming. Chemometr Intell Lab Syst 23:413–425
Ho TK (1998) The random subspace method for constructing decision forests. IEEE Trans Pattern Anal Mach Intell 20:832–844
Hsu CC, Chen MC, Chen LS (2010) A novel process monitoring approach with dynamic independent component analysis. Control Eng Pract 18:242–253
Hu KL, Yuan JQ (2008) Multivariate statistical process control based on multiway locality preserving projections. J Process Control 18:797–807
Hu Q, He Z, Zhang Z, Zi Y (2007) Fault diagnosis of rotating machinery based on improved wavelet package transform ans SVMs ensemble. Mech Syst Signal Process 21:688–705
Hwang DH, Han C (1999) Real-time monitoring for a process with multiple operating modes. Control Eng Pract 7:891–902
Hyvarinen A, Oja E (2000) Independent component analysis: algorithms and applications. Neural Netw, 13:411–430
Jarque CM, Bera AK (1987) A test for normality of observations and regression residuals. Int Stat Rev 55:163–172
Jin HD, Lee YH, Lee G, Han CH (2006) Robust recursive principal component analysis modeling for adaptive monitoring. Ind Eng Chem Res 45:696–703
Johannesmeyer MC, Singhal A, Seborg DE (2002) Pattern matching in historical data. AIChE J 48:2022–2038
Juricek BC, Seborg DE, Larimore WE (2001) Identification of the Tennessee Eastman challenge process with subspace methods. Control Eng Pract 9:1337–1351

Kano M, Nagao K, Hasebe H, Hashimoto I, Ohno H, Strauss R, Bakshi BR (2002) Comparison of multivariate statistical process monitoring methods with applications to the Eastman challenge problem. Comput Chem Eng 26:161–174

Kano M, Tanaka S, Hasebe S, Hashimoto I, Ohno H (2003) Monitoring independent components for fault detection. AIChE J 49:969–976

Kano M, Tanaka S, Hasebe S, Hashimoto I, Ohno H (2004a) Combined multivariate statistical process control. IFAC Symposium on advanced control of chemical processes (ADCHEM), Hong Kong, 11–14 January 2004, pp 303–308

Kano M, Hasebe S, Hashimoto I, Ohno H (2004b) Evolution of multivariate statistical process control: application of independent component analysis and external analysis. Comput Chem Eng 28:1157–1166

Kim D, Lee IB (2003) Process monitoring based on probabilistic PCA. Chemometr Intell Lab Syst 67:109–123

Kourti T (2005) Application of latent variable methods to process control and multivariate statistical process control in industry. Int J Adaptive Control Signal Process 19:213–246

Kramer MA (1992) Autoassociative neural networks. Comput Chem Eng 16:313–328

Kresta J, MacGregor JF, Marlin TE (1991) Multivariate statistical monitoring of process operating performance. Can J Chem Eng 69:35–47

Kruger U, Chen Q, Sandoz DJ, McFarlane RC (2001) Extended PLS approach for enhanced condition monitoring of industrial processes. AIChE J 47:2076–2091

Kruger U, Zhou YQ, Irwin GW (2004) Improved principal component monitoring of large-scale processes. J Process Control 14:879–888

Kruger U, Antory D, Hahn J, Irwin GW, McCullough G (2005) Introduction of a nonlinearity measure for principal component models. Comput Chem Eng 29:2355–2362

Kruger U, Kumar S, Littler T (2007) Improved principal component monitoring using the local approach. Automatica 43:1532–1542

Kruger U, Dimitriadis G (2008) Diagnosis of process faults in chemical systems using a local partial least squares approach. AIChE J 54:2581–2596

Krzanowski WJ (1979) Between-groups comparison of principal components. J Am Stat Assoc 74:703–707

Ku W, Storer RH, Georgakis C (1995) Disturbance detection and isolation by dynamic principal component analysis. Chemometr Intell Lab Syst 30:179–196

Kumar KV, Negi A (2008) SubXPCA and a generalized feature partitioning approach to principal component analysis. Pattern Recogn 41:1398–1409

Lee JM, Yoo CK, Choi SW, Vanrolleghem PA, Lee IB (2004) Nonlinear process monitoring using kernel principal component analysis. Chem Eng Sci 59:223–234

Lee JM, Yoo CK, Lee IB (2004a) Statistical process monitoring with independent component analysis. J Process Control 14:467–485

Lee JM, Yoo CK, Lee IB (2004b) Statistical monitoring of dynamic processes based on dynamic independent component analysis. Chem Eng Sci 59:2995–3006

Lee JM, Qin SJ, Lee IB (2006a) Fault detection and diagnosis based on modified independent component analysis. AIChE J 52:3501–3514

Lee YH, Jin HD, Han CH (2006b) On-line process state classification for adaptive monitoring. Ind Eng Chem Res 45:3095–3107

Li RY, Rong G (2006) Fault isolation by partial dynamic principal component analysis in dynamic process. Chin J Chem Eng 14:486–493

Li RF, Wang XZ (2002) Dimension reduction of process dynamic trends using independent component analysis. Comput Chem Eng 26:467–473

Li W, Qin SJ (2001) Consistent dynamic PCA based on error-in-variable subspace identification. J Process Control 11:661–676

Li W, Yue HH, Valle-Cervantes S, Qin SJ (2000) Recursive PCA for adaptive process monitoring. J Process Control 10:471–486

Lieftucht D, Kruger U, Irwin GW (2006) Improved reliability in diagnosing faults using multivariate statistics. Comput Chem Eng 30:901–912

Lieftuch D, Kruger U, Irwin GW, Treasure RJ (2006) Fault reconstruction in linear dynamic systems using multivariate statistics. IEE Proc Control Theory Appl 153:437–446

Liu XQ, Xie L, Kruger U, Littler T, Wang SQ (2008) Statistical-based monitoring of multivariate non-Gaussian systems. AIChE J 54:2379–2391

Lyman PR, Georgakist C (1995) Plant-wide control of the Tennessee Eastman problem. Comput Chem Eng 19:321–331

MacGregor JF, Jaeckle C, Kiparissides C, Koutoudi M (1994) Process monitoring and diagnosis by multiblock PLS methods. AIChE J 40:826–838

Malhi A, Gao R (2004) PCA-based feature selection scheme for machine defect classification. IEEE Trans Instrum Meas 53:1517–1525

Maulud A, Wang D, Romagnoli JA (2006) A multi-scale orthogonal nonlinear strategy for multivariate statistical process monitoring. J Process Control 16:671–683

Mika S, Scholkopf B, Smola A, Muller KR, Scholz M, Ratsch G (1998) Kernel PCA and de-noising in feature spaces. MIT Press, Denver, pp 536–542

Miller P, Swanson RE, Heckler CF (1993) Contribution plots: the missing link in multivariate quality control. In: Fall Conference of the ASQC and ASA, Milwaukee, WI

Narasimhan S, Shah S (2008) Model identification and error covariance matrix estimation from noisy data using PCA. Control Eng Pract 16:146–155

Nomikos P, MacGregor JF (1994) Monitoring batch processes using multiway principal component analysis. AIChE J 44:1361–1375

Nomikos P, MacGregor JF (1995a) Multivariate SPC charts for monitoring batch process. Technometrics 37:41–59

Nomikos P, MacGregor JF (1995b) Multi-way partial least square in monitoring batch processes. Chemometr Intell Lab Syst 30:97–108

Nounou MN, Bakshi BR, Goel PK, Shen XT (2002) Bayesian principal component analysis. J Chemometr 16:576–595

Odiowe PP, Cao Y (2010) State-space independent component analysis for nonlinear dynamic process monitoring. Chemometr Intell Lab Syst 103:59–65

Overschee PV, Moor DM (1996) Subspace identification for linear systems. Kluwer Academic Publishers, Boston

Qin SJ (1998) Recursive PLS algorithms for adaptive data monitoring. Comput Chem Eng 22:503–514

Qin SJ (2003) Statistical process monitoring: basics and beyond. J Chemometr 17:480–502

Qin SJ, Li WH (2001) Detection and identification of faulty sensors in dynamic processes. AIChE J 47:1581–1593

Qin SJ, Valle S, Piovoso MJ (2001) On unifying multiblock analysis with application to decentralized process monitoring. J Chemometr 15:715–742

Raich AC, Çinar A (1995) Multivariate statistical methods for monitoring continuous processes: assessment of discrimination power of disturbance models and diagnosis of multiple disturbances. Chemometr Intell Lab Syst 30:37–48

Raich A, Cinar A (1997) Diagnosis of process disturbance by statistical distance and angle measures. Comput Chem Eng 21:661

Ricker NL (1996) Decentralized control of the Tennessee Eastman challenge process. J Process Control 6:205–221

Schölkopf B, Smola AJ, Müller K (1998) Nonlinear component analysis as a kernel eigenvalue problem. Neural Comput 10:1000–1016

Sebzalli YM, Wang XZ (2001) Knowledge discovery from process operational data using PCA and fuzzy clustering. Eng Appl Artif Intell 14:607–616

Simoglou A, Georgieva P, Martin EB, Morris AJ, Feyo de Azevedo S (2005) On-line monitoring of a sugar crystallization process. Comput Chem Eng 29:1411–1422

Singhal A, Seborg DE (2002) Pattern matching in multivariate time series databases using a window-moving approach. Ind Eng Chem Res 41:3822–3838

Singhal A, Seborg DE (2006) Evaluation of a pattern matching method for the Tennessee Eastman challenge process. J Process Control 16:601–613

Smilde AK, Westerhuis JA, de Jong S (2003) A framework for sequential multiblock component methods. J Chemometr 17:323–337

Srinivasan R, Wang C, Ho WK, Lim KW (2004) Dynamic principal component analysis based methodology for clustering process states in agile chemical plants. Ind Eng Chem Res 43:2123–2139

Stander CJ, Heyns PS, Schoombie W (2002) Using vibration monitoring for local fault detection on gears operating under fluctuating load conditions. Mech Syst Signal Process 16:1005–1024

Staszewski WJ, Tomlinson GR (1997) Time-frequency analysis in gearbox fault detection using the Wigner-Ville distribution and pattern recognition. Mech Syst Signal Process 11:673–692

Stefatos G, Ben Hamza A (2010) Dynamic independent component analysis approach for fault detection and diagnosis. Expert Syst Appl 37:8606–8617

Stork CL, Veltkamp DJ, Kowalski BR (1997) Identification of multiple sensor disturbances during process monitoring. Anal Chem 69:5031–5036

Suykens JAK, Van Gestel T, De Brabanter J, De Moor B, Vandewalle J (2002) Least squares support vector machines. World Scientific, Singapore

Takahashi T, Kurita T (2002) Robust de-noising by kernel PCA. Springer, Berlin, pp 739–744

Tan CK, Mba D (2005) Limitation of acoustic emission for identifying seeded defects in gearboxes. J Nondestruct Eval 24:11–28

Tax DMJ, Duin RPW (1999) Support vector domain description. Pattern Recongnit Lett 20:1191–1199

Tax DMJ, Duin RPW (2004) Support vector domain description. Mach Learn 54:45–66

Thissen U, Swierenga S, de Weijer AP, Wehrens R, Melssen WJ, Buydens LMC (2005) Multivariate statistical process control using mixture modeling. J Chemometr 19:23–31

Thornhill NF, Horch A (2007) Advances and new directions in plant-wide disturbance detection and diagnosis. Control Eng Prac 15:1196–1206

Tipping ME, Bishop CM (1999a) Probabilistic principal component analysis. J Roy Stat Soc 61:611–622

Tipping ME, Bishop CM (1999b) Mixtures of probabilistic principal component analysis. Neural Comput 11:443–482

Valle S, Li WH, Qin SJ (1999) Selection of the number of principal components: the variance of the reconstruction error criterion with a comparison to other methods. Ind Eng Chem Res 38:4389–4410

Vapnik VN (1995) The nature of statistical learning theory. Springer, New York

Venkatasubramanian V, Rengaswamy R, Kavuri SN, Yin K (2003) A review of process fault detection and diagnosis: Part III: process history based methods. Comput Chem Eng 27:327–346

Vincenzo N, Guiseppe Q, Aniello F (2008) The detection of gear noise computed by integrating the Fourier and wavelet methods. WSEAS Trans Signal Process 3:60–67

Wang W (2008) Autoregressive model-based diagnostics for gears and bearings. Insight Non Destr Test Cond Monit 50:414–418

Wang HQ, Song ZH, Li P (2002) Fault detection behavior and performance analysis of PCA-based process monitoring methods. Ind Eng Chem Res 41:2455–2464

Wang X, Kruger U, Lennox B (2003) Recursive partial least squares algorithms for monitoring complex industrial processes. Control Eng Pract 11:613–632

Wang HQ, Zhou HL, Hang BL (2004) Number selection of principal components with optimized process monitoring performance. IEEE Conf Decis Control 5:4726–4731

Wang XZ, Medasani S, Marhoon F, Albazzaz H (2004) Multidimensional visualization of principal component scores for process historical data analysis. Ind Eng Chem Res 43:7036–7048

Wang X, Kruger U, Irwin GW (2005) Process monitoring approach using fast moving window PCA. Ind Eng Chem Res 44:5691–5702

Wang X, Kruger U, Irwin GW, McCullough G, McDowell N (2007) Nonlinear PCA with the local approach for diesel engine fault detection and diagnosis. IEEE Trans Control Syst Technol 16:122–129

Westerhuis JA, Kourti T, MacGregor JF (1998) Analysis of multiblock and hierarchical PCA and PLS models. J Chemometr 12:301–321

Westerhuis JA, Gurden SP, Smilde AK (2000) Generalized contribution plots in multivariate statistical process monitoring. Chemometr Intell Lab Syst 51:95–114

Widodo A, Yang BS (2007) Support vector machine in machine condition monitoring and fault diagnosis. Mech Syst Signal Process 21:2560–2574

Widodo A, Yang BS, Han T (2007) Combination of independent component analysis and support vector machines for intelligent faults diagnosis of induction motors. Expert Sys Applications 32:299–312

Wu JD, Hsu CC, Wu GZ (2008) Fault gear identification and classification using discrete wavelet transform and adaptive neuro-fuzzy inference. Expert Syst Appl 36:6244–6255

Xie L, Wu J (2006) Global optimal ICA and its application in MEG data analysis. Neurocomputing 69:2438–2442

Xie L, Kruger U, Lieftucht D, Littler T, Chen Q, Wang SQ (2006) Statistical monitoring of dynamic multivariate processes part1 modeling autocorrelation and cross-correlation. Ind Eng Chem Res 45:1659–1676

Yao Y, Gao F (2008) Subspace Identification for Two-Dimensional Dynamic Batch Process Statistical Monitoring. Chem Eng Sci 63:3411–3418

Yao Y, Gao FR (2009) A survey on multistage/multiphase statistical modeling methods for batch processes. Annu Rev Control 33:172–183

Yu J, Qin SJ (2008) Multimode process monitoring with Bayesian inference-based finite Gaussian mixture models. AIChE J 54:1811–1829

Yu J, Qin SJ (2009) Multiway Gaussian mixture model based multiphase batch process monitoring. Ind Eng Chem Res 48:8585–8594

Yue HH, Qin SJ (2001) Reconstruction-based fault identification using a combined index. Ind Eng Chem Res 40:4403–4414

Zhang YW, Qin SJ (2007) Fault detection of nonlinear processes using multiway kernel independent component analysis. Ind Eng Chem Res 46:7780–7787

Zhang YW, Qin SJ (2008) Improved nonlinear fault detection technique and statistical analysis. AIChE J 54:3207–3220

Zhang YW, Zhang Y (2010) Fault detection of non-Gaussian processes based on modified independent component analysis. Chem Eng Sci 65:4630–4639

Zhang QH, Basseville M, Benveniste A (1994) Early warning of slight changes in systems and plants with application to condition based maintenance. Automatica 30:95–114

Zhang J, Martin EB, Morris AJ (1996) Fault detection and diagnosis using multivariate statistical techniques. Chem Eng Res Design 74:89–96

Zhang MG, Ge ZQ, Song ZH (2010). Probabilistic process monitoring with Bayesian regularization. American Control Conference—ACC2010, Baltimore, Maryland, pp 6999–7003

Zhao SJ, Xu YM (2005) Multivariate statistical process monitoring using robust nonlinear principal component analysis. Tsinghua Sci Technol 10:582–586

Zhao SJ, Zhang J, Xu YM (2004) Monitoring of processes with multiple operation modes through multiple principle component analysis models. Ind Eng Chem Res 43:7025–7035

Zhao CH, Wang FL, Jia MX (2007) Dissimilarity analysis based batch process monitoring using moving windows. AIChE J 53:1267–1277

Zhao CH, Wang FL, Mao ZZ, Lu NY, Jia MX (2008a) Adaptive monitoring based on independent component analysis for multiphase batch processes with limited modeling data. Ind Eng Chem Res 47:3104–3113

Zhao CH, Wang FL, Gao FR, Zhang YW (2008b) Enhanced process comprehension and statistical analysis for slow-varying batch processes. Ind Eng Chem Res 47:9996–10008

Zhao CH, Wang FL, Zhang YW (2009) Nonlinear process monitoring based on kernel dissimilarity analysis. Control Eng Pract 17:221–230

Index

A
Adaptive partial least squares (PLS), 82, 95
Adaptive principal component analysis (PCA), 2, 82, 95
Advanced fault identification, 170, 172, 173, 175, 178, 180, 181
Akaike Information Criterion, 14
Algorithm complexity analysis, 79, 80, 121
Auto-correlation, 134, 136, 138, 142

B
Backward elimination sensor identification, 29
Bayesian combination, 114
Bayesian inference, 2, 3, 62, 66, 69, 78, 96, 99, 101, 102, 111, 117, 118, 129
Bayesian Information Criterion, 14
Bayesian principal component analysis, 147, 150, 152–154, 166
Bayesian regularization, 3, 147, 148, 150, 152, 158, 165
Block contributions, 173

C
Confidence limit, 6, 14, 20, 25, 26, 50, 56, 70, 142, 143, 173, 177
Contribution plot, 30, 62, 169, 170, 172, 173, 175, 177, 180
Critical fault magnitude, 77
Cross-correlation, 9, 134, 136, 138, 142

D
Detectability, 29, 62
Dimension reduction, 10, 80, 96, 97
Dynamic independent component analysis (DICA), 132, 139, 142, 146
Dynamic process monitoring, 2, 131
Dynamic statistical-based approach, 147

E
Essential variable, 13, 14, 22, 24
Expectation-maximization, 149
External analysis, 96, 113, 114

F
Factor analysis, 1–3, 7, 14
Fast independent component analysis (FastICA), 23, 138
Fault detection, 3, 14, 16, 20, 24, 26, 29, 100, 126, 171, 172, 175, 179–181
Fault detection index, 119
Fault diagnosis, 3, 29, 32, 42, 67, 69, 71, 78, 171
Fault identification, 3, 30, 36, 40, 96, 98, 101, 173, 180
Fault pool, 129
Fault probability, 62, 66, 73, 79, 96
Fault reconstruction, 30, 175
Fault scenario, 75, 96, 101, 111
Feature extraction, 95–97, 103, 111
Finite impulse response, 81

G
Gaussian, 7, 10, 24, 61, 81, 96, 113, 177
Gaussian mixture model, 96, 102, 113, 122
Gearbox system, 131–133, 141, 146
Global contributions, 175

I
Improved residual, 50–52, 58, 140, 141, 145
Incipient faults, 46, 58, 131, 141, 143, 146
Independent component, 10, 14, 26, 30, 38, 89, 90, 98, 103, 107, 137, 171
Independent component analysis, 1, 9, 13, 29, 82, 95, 132, 137, 170

Z. Ge, Z. Song, *Multivariate Statistical Process Control*, Advances in Industrial Control,
DOI 10.1007/978-1-4471-4513-4, © Springer-Verlag London 2013

Independent component subspace, 169
Isolatability, 62

J
Joint probability, 103, 105–109, 111, 156, 157, 165
Just-in-time-learning, 81

K
Kalman innovation, 133, 138, 139, 141, 142, 146
Kernel density estimation, 14, 99, 173
Kernel principal component analysis, 1–3, 31, 45, 61, 81, 113, 121, 123

L
Latent factor, 7
Linear correlation, 116
Linear subspace, 2, 3, 62, 63, 65, 78, 114–120
Local approach, 3, 46, 47, 50, 53, 56, 58, 132
Local Bayesian principal component analysis, 152–154
Local least squares support vector regression, 3
Local model, 82, 84–89, 92–94, 96

M
Mahalanobis distance, 25
Main angle, 38
Maximum-likelihood, 7, 153
Maximum-likelihood principal component analysis, 7, 8, 23, 149, 153
Mixture Bayesian regularization, 148, 152, 154, 165
Mode clustering, 96, 97, 120
Mode identification, 40, 42
Mode localization, 129, 148, 156, 159, 160, 164, 165
Model library, 82, 96
Multiblock principal component analysis, 2, 170, 179, 181
Multidimensional faults, 29, 30, 36, 174
Multimode, 2–4, 95, 96, 98, 102, 106, 113, 114, 122, 123, 128, 148
Multimode process monitoring, 2, 3, 96, 98, 113, 152, 154, 165
Multivariate statistical process control, 11, 45, 61, 81, 95, 113, 131, 169

N
Neural network, 2, 45, 61, 81, 83, 84, 113
Non-Gaussian, 27, 38, 39, 42, 88, 97, 113, 132, 133, 136, 138, 143, 146, 170–172, 177
Non-Gaussian fault reconstruction, 43
Non-Gaussian process monitoring, 2, 27
Nonlinear fault detection index, 119
Nonlinear multimode process monitoring, 3, 113, 114, 128
Nonlinear process monitoring, 2, 3, 45, 46, 49, 58, 61, 65, 78–81, 113, 128
Normal operating condition, 16, 51

P
Partial least squares, 1, 6, 29, 81, 95, 113, 147, 169
Plant-wide process monitoring, 2, 3, 170, 176, 181
Posterior probability, 96, 100, 101, 103, 109, 111, 118, 129, 148, 153–157, 166
Primary residual, 47–50, 140
Principal component, 1, 36, 70, 75, 80, 90, 98, 103, 147–149, 154, 162, 165, 171,
Principal component analysis, 5, 13, 29, 30, 45, 48, 61, 95, 113, 116, 141, 147, 169
Principal component subspace, 63, 169
Principal curve, 45, 61, 81, 113
Prior probability, 66, 100, 118, 155
Probabilistic principal component analysis, 1–3, 7, 14
Probabilistic process monitoring, 2, 148,
Process monitoring, 1–3, 6, 8, 11, 14, 25, 45, 47, 49, 53, 61, 69, 95, 106, 147, 162, 169, 171, 179

R
Reconstructability, 62
Reconstruction-based contribution, 62, 67
Recursive principal component analysis, 82
Residual subspace, 38, 63, 169
Responsible sub-block, 170, 172, 173, 175, 179

S
Similarity factor, 30, 36, 38, 39, 41–43, 87
Singular vector decomposition, 174
Statistical analysis, 128
Statistical process control, 1, 131
Subspace contribution index, 64, 70
Subspace identification, 140, 142
Subspace model identification based local approach, 140, 143, 145, 146
Support vector data description, 1, 14, 21, 30, 132
Support vector regression, 3, 81

T
Tennessee Eastman, 17, 52, 122, 175
Time-varying process monitoring, 2, 82, 88

Two-dimensional Bayesian, 114, 117, 121, 128
Two-level multivariate statistical monitoring, 170, 171, 179, 181, 182
Two-step fault identification, 182

U
Unidimensional fault, 29, 174
Univariate monitoring statistic, 30, 139–142, 146

MIX
Papier aus verantwortungsvollen Quellen
Paper from responsible sources
FSC® C105338

If you have any concerns about our products,
you can contact us on
ProductSafety@springernature.com

In case Publisher is established outside the EU,
the EU authorized representative is:
Springer Nature Customer Service Center GmbH
Europaplatz 3, 69115 Heidelberg, Germany

Printed by Libri Plureos GmbH
in Hamburg, Germany